The Platform Economy

A different version of chapter 3 was previously published as "Genesis of the Platform Concept: I-mode and Platform Theory in Japan," *Asiascape: Digital Asia* 4, no. 3 (2017): 184–208. A different version of chapter 5 was previously published as "Converging Contents and Platforms: Niconico Video and Japan's Media Mix Ecology," in *Asian Video Cultures,* ed. Joshua Neves and Bhaskar Sarkar, 91–113 (Durham, N.C.: Duke University Press, 2017). The author thanks the publishers for their permission to reprint this material.

Published by the University of Minnesota Press
111 Third Avenue South, Suite 290
Minneapolis, MN 55401-2520
http://www.upress.umn.edu

ISBN 978-1-5179-0694-8 (hc)
ISBN 978-1-5179-0695-5 (pb)

A Cataloging-in-Publication record for this book is available from the Library of Congress.

Printed in the United States of America on acid-free paper

The University of Minnesota is an equal-opportunity educator and employer.

UMP BmB 2019

Contents

Platform Worlds

Platforms are everywhere. As digital objects we have social media platforms (Facebook, Twitter, Weibo), chat apps as platforms (LINE, Messenger, WhatsApp), e-commerce platforms (Amazon, Alibaba, Rakuten, Flipkart), streaming platforms (Netflix, Mubi, Niconico Video, Youku), and smartphones (iOS and Android). As places we have bookstores as platforms, storefronts as platforms. We have educational platforms, political platforms, business platforms. We have gone from the era of platform shoes as a distinct genre of footwear to platform everything. And this list does not even scratch the list of what is called a platform, or what is retroactively redescribed as one. The greatest success of *platform* within our language ecosystem is to have become something of a universal translation device. Almost anything can become a platform, if one merely calls it such. And thereupon it immediately gains prestige, cool, and economic value. "Platforms are eating our world," as one business book on the topic succinctly put it.[1] To which we should add a caveat: platforms are as much technical objects and business practices as they are a keyword—with Exhibit A being the now-ubiquitous smartphone.

If keywords become the mediating nodes for the redescription of worlds—aggregators of currency, thought, conversation, cultural value, and labor—platform is without doubt the keyword of our time. We live in a platform world. And this "we" is not only the Anglophone linguistic subject. The term is remarkable for its portability to other languages in its numerous transliterations: we have *purattofōmu* in Japanese or *peullaes-pom* in Korea, *plataformas* in Spanish, *plateforme* in French (possibly the root of the term), *píngtái* (平台) or "flat platform/stage" in Chinese. The world is in the midst of a "platform revolution," to borrow the title of one recent tome, and a shift to "platform capitalism" in the formulation of another.[2] Academic discourse is as afflicted as other spheres in this count, with *platform* becoming the stand-in term for any media or device after the digital shift. *Contents (kontentsu)*, a term that comes from a similar set of circumstances but a different lineage, is its partner term on the rise to linguistic and strategic dominance (rendered here as contents, reflecting the always-pluralized Japanese version of the word that I mostly follow here, unless explicitly referring to the non-Japanese context, or quoting from an English translation of Japanese that renders the term as *content*). Platform and contents organize industries, government strategies, and popular understandings of media. They are the keywords of our time. They have also become—true to the colonizing drive embedded in these terms—the keywords for all times, with Shakespeare rebranded as a content creator, and bookstores retroactively termed *platforms*.[3]

The Platform Economy offers a systematic examination of the term *platform* from *historical, geographical, cultural, institutional, and corporate* perspectives, providing a synthetic account of its discursive development and practical uses. The "economy" of the title is meant to designate both the transformation of the economy proper—its increased dependence on digital platforms and the emergence of the platform economy—and the metaphoric economy and circulation of the term *platform* as a keyword, which crosses domains and informs cultural and industrial forms of production. This book gives an overview of platform theory, traces the discursive and general economy of platform as a keyword for the media industries, and examines how the concept was unfolded in practice in the development of Japan's mobile internet ecosystem, its *keitai*, or feature phones. It was Japan's mobile internet project that became the blueprint for the now-dominant mobile "platforms" of Apple's iPhone/iOS and Google's Android phones—technical objects, to be sure, but more importantly points of transaction or interface that support contemporary capitalist forms of accumulation. This book tracks this and also the long-term impact of Japan's mobile platforms on the Japanese internet, the regional impacts in East Asia, and the transformation of the commercial internet writ large via the smartphone (or smartphonization of the internet, which bears the imprint of the Japanese feature phone legacy). While in Japan the mobile internet and the PC-based web are regarded as completely different beasts—to the extent that a deep dive into internet history can focus solely on the PC web and address the mobile internet only in conclusion—this book treats the mobile internet as a platform prototype for contemporary internet logics full stop.[4] This book presents an analysis of the unfolding of platform that follows it from theoretical construct, to keyword, to planning, to product, to the lasting cultural and economic formation that it is today—from platform terminology to platform economy, as it were. Platforms are much more than just words, then; they are also technological and managerial constructs that mediate our relationship to our worlds, that create habits, addictions, and impulses (like the drive to check notifications), and just generally vie for our attention and shape our lives. Platforms designate not just particular technological entities but managerial constructs that shape us and the relations we enter into with other people, companies, and objects. This book offers one account of the rise of the platform as economic formation and manager of our lives.[5]

From Words to Worlds

Keywords matter. They operate on the world, they model the world, and they inscribe a series of operations (technical, strategic, financial) that transform the world. Never a deterministic process, the relation between words and worlds, or discourse and things, or theory and practice, is open, recursive, and dynamic. Patterns can be observed; regularities in discourse and regularities in practice can imprint one on the other.[6] Platform theory leads to platform-building practice. Practice then acts back on this theory, leading to the ascendancy of, for instance, a transactional model of the platform that is central to contemporary economic forms of this object. *Contents* is similarly operationalized as a term, enacting a form of capture, a magical and even alchemic transformation of data or information into monetizable things.

Maurizzio Lazzarato has argued that contemporary capitalism is characterized not so much by the *creation of products* but by the *creation of worlds*. In his 2004 book, *Les revolutions du capitalisme* (The revolutions of capitalism), Lazzarato writes that the contemporary enterprise "creates not the object (the merchandise) but the world where the object exists. It creates not the subject (worker or consumer) but the world where the subject exists."[7] Capitalist valorization thus depends on the development of worlds. Lazzarato continues:

> In reversing the Marxist definition, we can say that capitalism is not a mode of production but a production of modes, a production of worlds *[une production de mo(n)des].* . . . The expression and the effectuation of worlds and of the subjectivities which are included therein, and the creation and realization of the sensible . . . precede economic production."[8]

In this passage Lazzarato productively fuses *monde* (world) and *mode* (mode), relating capitalist *modes of production* to the *production of worlds*. In what follows I do to words what Lazzarato does to modes; I argue that capitalism and language cooperate in the production of wor(l)ds: words and worlds. Words operate on, model, and cocreate the business ideologies, strategies, and mechanisms that shape our world. To paraphrase Lazzarato: capitalist valorization depends on the development of words–keywords, like *contents* and *platforms*.

We can find keywords' transformation of our world at work in financial markets, in CEO letters to shareholders, in tech writing, and in business literature in general.[9] A crucial site from which words operate on worlds is in and through management theory, whose contributing writers come from a variety of disciplines, including management studies, economics, and organizational theory. Management theory emerges from what Walter Kiechel in his book on the management consulting industry calls an impetus toward "greater Taylorism"—the expansion of the Taylorist idea of rationalizing production from the production process narrowly defined to the entire firm or corporation.[10] As Kiechel writes: "The secret intellectual history of the new corporate world is as much about the challenges companies faced, from competing with the Japanese in the 1970s to surviving a crisis in the global financial system in the twenty-first century, as it is about the conceptual solutions devised in response."[11] Management theory and management consulting are two adjoining sites where crisis meets proposed solution, and where language and analysis inaugurates a new era of corporate strategy and practice. While Kiechel focuses on the history of management consulting, *The Platform Economy* examines the near history of a set of keywords, discourses, and corporate practices that take shape in the 1990s and come to dominance in the mid-2010s in the form of what some have termed *platform capitalism,* and what I will refer to as the *platform economy.*[12] Management theory, government white papers, and technology writing are just some of the key sites from which this paradigm emerges.

It is from within business literature and management theory—or what we might call "greater management theory," which includes professional consultants as well as academic economists, many of whom write popular books and have a consulting practice on the side—that we find the development of one of the main strands of platform theory that this book relies on: the model of platform as intermediary that facilitates third-party transactions.[13] It is this model that informs many contemporary businesses that call themselves platforms today—from Uber to Amazon to YouTube. For this reason, management work and business literature are indispensable sources for developing a critical analysis and history of platform capitalism. Management theory is, in turn, buttressed by practices of platform building by corporations I examine in this book. Platform capitalism and the platform economy hence depend as much on the language around platforms (platform discourse) as the practical enactments of platforms (platform building). This account contrasts with the strong emphasis on platform-as-technology found within many books on the subject, particularly those within Ian Bogost and Nick Montfort's Platform Studies book series, as well as the general turn to infrastructure studies within media studies. It aligns itself rather with work that attends to the vocabulary of digital media, "cultural logic of computation," as David Golumbia puts it, or the punctuation of the digital shift to which Jeff Scheible calls our attention.[14] Indeed, much as Golumbia sets out to "question not the development of computers themselves but the emphasis on computers and computation that is widespread throughout almost every part of the social fabric,"[15] this book sets out to narrate the platform as rhetorical device, as well as object. It also aims to narrate the transformative powers platforms and the conceptual models behind them have exerted on our world.[16]

In one of the earliest critiques of the terminological rise of platform, Tarleton Gillespie notes that "the discourse of the 'platform' works against us developing such [linguistic or terminological] precision, offering as it does a comforting sense of technical neutrality and progressive openness."[17] Indeed, as Gillespie convincingly argues, platform as keyword *operates* so effectively precisely because of the discursive fudge it performs, encompassing progressively more objects until capitalism itself is no longer *late* or *post-Fordist* but rather *platform* in kind—as Nick Srnicek's *Platform Capitalism* suggests. No doubt, *platform capitalism* is a useful descriptive term for the centrality that platform companies—such as Alphabet (Google), Facebook, Alibaba, Amazon, Tencent, SoftBank, Netflix, and Flipkart—have in the construction of the contemporary form of capitalism. Platforms are even the mode of interface to the financial sector, wherein financial capital may itself be a subset of platform capitalism (dependent as it is on the one hand on digital infrastructure for its effectuation of trades, and on the other hand on the high valuation of platform companies for seeking return on investment). However, while platform capitalism is ultimately a useful critical formulation—one that I deploy here as a subset of post-Fordism as periodizing term—the very rise of the term *platform capitalism* should alert us to the very terminological expansionism that we are facing, and that this book narrates. Hence *The Platform Economy* is the title and operation of this book, which examines the metaphorical economy of terminology as it moves through and reorganizes fields, as much as the effects on the financial economy. The origins of the term *economy* in *oikonomia* or the *management* of a household further suits this book's emphasis on the patterns of management (rhetorical and operational) that create the conditions for platform capitalism. It is the wider economy of platform as term and practice of management that is the object here.

This book takes stock of what a platform is today, and how it has become such. It offers a three-fold typology of the platform, defined as (1) a layered structure often based on hardware, (2) a support for contents, and (3) a structure of mediation or enabler of financial transactions. It examines each in turn and emphasizes that most platforms are hybrids of the above. This book emphasizes history over typology, however, offering a recent history of the term *platform* and the practices associated with platform design, focusing on a time period near at hand but also light years away in internet time, beginning in the 1990s, and examining the following decade, before arriving at the present moment. It also tells the wider story of the emergence of the smartphone, and of how we came to pay for media via our phones and on the internet—a significant transformation in cultural attitudes and day-to-day media practices—via the Japanese story of platforms. In so doing, *The Platform Economy* examines the effects of platform-speak on media industries and their production and monetization of contents. It interrogates the discursive parameters and industry impacts of *platform* as keyword and as thing, weaving together an account of both platform theory and platform development in Japan. In doing so it also picks up on the ways that Japanese hardware such as automobiles, VCRs, and game consoles in turn informed work in business literature that leads, ultimately, to platform theory as we know it today. Japan, this book will claim, is a key site for the development of both contemporary platforms *and* platform theory. It is also a site from which the platformization of the internet can be seen most clearly, both in management theory and in practice. The principle guiding this book is that a firm critique of platform capitalism must be grounded in a better grasp of the development of platforms and platform theory in their historical and geographical diversity.

From Network to Platform

This is also, implicitly, the story of the displacement of *network* by *platform*. From the 1990s until at least the middle of the following decade, *network* was the keyword of the time.[18] Coined in the early 1990s, "network society" was the phrase that described the social formation of the world connected by the internet, as described by prolific writers such as Manuel Castells in books such as *The Rise of the Network Society* (1996). Castells was writing from the perspective of a critical sociology. But the term *network* was equally operative within management literature of the 1990s, as detailed by Luc Boltanski and Eve Chiapello in *The New Spirit of Capitalism*. As Boltanski and Chiapello note, "The *metaphor of the network* . . . is deployed in all sorts of contexts" and becomes a master metaphor in management literature of the 1990s.[19] For them (and I am in complete agreement with their assessment), this is significant, as "management literature is a medium offering the most direct access to the representations associated with the spirit of capitalism in a given era"; it is also "one of the main sites in which the spirit of capitalism is inscribed."[20] Not merely representational, management literature is also operational, however: it is the site of description but also the site of inscription of ways of thinking about the world. Taken up by corporate actors, management theory is transformed into an applied theory, operating on and transforming the world.

What *network* was for the 1990s and the following decade, *platform* is for the mid-2010s onward. As Geert Lovink rightly remarks, 2016 marks the year of platform's official ascendance to the throne of Keyword of Our Time.[21] A couple of key events might be enumerated to explain why he suggests 2016 is the turning point. The year saw the publication of the two most significant popular management books on the topic—*Platform Revolution: How Networked Markets Are Transforming the Economy and How to Make Them Work for You* and *Matchmakers: The New Economics of Multisided Platforms* by a group of economists and management writers who had been pushing the term in their field for the previous decade (in both academic and public-facing writing).[22] It was also the year of the Platform Society conference held at the Oxford Internet Institute, followed by two journal special issues on the topic. Srnicek's *Platform Capitalism* similarly arrived on the scene in 2016, further solidifying the case that 2016 was the year platform officially became the keyword of our time.[23] Of course, despite the later debut of *platform* as a keyword in the first decade of the twenty-first century, and its coronation in 2016 with the declaration that we now exist within a platform society, economy, or capitalism, the argument in *The Platform Economy* is that management literature had been treating the term *platform* with considerable seriousness since at least the mid-1990s, making it congruent to the heyday of *network* as keyword—within managerial literature and outside it—even if it took a little longer for the term to catch on.

As they figure within management literature, the terms *platform* and *contents* serve to manage and effectuate the digital shift as much as they serve to describe it. In this sense we may recall Alan Liu's critical call to analyze management discourse as a site of the management of the transition from industrialism to postindustrialism. Liu writes:

> A cultural studies approach might show that the alignment of the new discourse with our new economy is the result of a historical process of *making* things fit. That process we now call management, the modern theory of civilization. God begat Enlightenment reason, which begat industrial scientific management, which in turn begat postindustrial management theories that synonymize the progress of civilization and management without any remainder.[24]

The terms *platform* and *contents,* and the swirling, voluminous density of business and popular writing around them, suggest that the keywords are operationalized to, as Liu says, *make things fit.* After all, as Boltanski and Chiapello argue, management theory and its keywords do not just represent the new spirit of capitalism; they also effectuate it. Ultimately, the rise of the conceptual model of platform as universal mediation device within management theory and its effectuation in actual platform practice marks the infiltration of managerial logics more fully into the social body, producing what we now call the platform society.

Platforms in Japan

To better understand this effectuation, *The Platform Economy* departs from the American context that is usually figured as the homeland of platform capitalism. Against the still overwhelming emphasis on U.S. platforms within popular and academic writing, and also contrary to the treatment of the platform in a resolutely presentist mode, this book will narrate the what, where, and how of the platform economy by way of a focus on Japan—a fading platform superpower. There are other such fading superpowers, such as France, in the parallel case of its Minitel system, which Julien Mailland and Kevin Driscoll have recently argued offers an early formulation of the platform concept.[25] That said, by other measures, Japan makes a good case for it being a platform progenitor. It generated some of the crucial platforms we refer to, whether these are video cassette recorders in the late 1970s, Nintendo Entertainment Systems in the 1980s and 1990s, or the mobile internet system called i-mode in the late 1990s and the following decade. These platforms in turn impact the Japanese internet ecology of the early twenty-first century, and the smartphone experience globally. The very conceptual framework of the "platform business" is closely related to the structure of payment for contents implemented with the development and rollout of the world's first widespread mobile internet platform—i-mode, released by telecom giant NTT Docomo in February 1999, becoming what one writer called "the biggest phenomenon since the Walkman."[26] I-mode inaugurated and formatted a mobile internet experience predicated on services that became a model for the now-dominant mobile operating systems, Google's Android and Apple's iOS. These are platforms and sources of monetization for their respective companies, which in turn transformed the internet from the venture capital–supported, free services model of the web to the mobile internet as the app-based and increasingly fee-supported model (whether these are for paid apps, or in-app purchases). To be sure, the ad-supported, eyeball-and-data-seeking "free internet" ethos of the web persists, but it does so in coexistence with a paid model that was pioneered in the mobile realm in the late 1990s in Japan.

That for which we pay goes by the moniker *contents*—as significant a keyword as its partner term *platform*. Contents is hence also part of the narrative here, and indeed one of the particularities of Japanese platforms has been the closer integration of contents into the platforms, and the distinct interfaces that this integration gives rise to. The pervasiveness of the term *contents* (again, always in the plural form in Japan) in fact precedes that of platforms. The term has often been tied to distinct types of media in Japan, such as anime and manga and video games, before it became tied to mobile media. Contents exists in close relation to platforms. There is, as Thomas Lamarre observes, "a generative interval . . . between platform and content" that supports a market for—in the context he is writing—home video games and their consoles.[27] By the early twenty-first century, the mobile contents industry in Japan exploded into view with internet-enabled mobile phones; the mobile contents market is huge and still growing.[28] It was one of the reasons the mobile internet caught on at all.

The financial success and cultural impact of the mobile contents market was in turn one of the reasons it became a keyword, eventually finding its way into "Cool Japan" national policy-making, which attempts (often rather crudely and ineffectively) to capitalize on the global popularity and exportability of Japanese animation, character goods, and eventually food and cosmetics brands.[29] Despite the continuing uptick in the market value for mobile contents, however, the rise of platforms is also a moment that sees the decline in the discursive weight and import of contents. Consequently, we live within not *contents capitalism* but *platform capitalism,* and today business strategists urge us to avoid the "content trap," relying instead on the connectivity that these contents make possible—in other words, business owners are urged to focus on connective platforms rather than contents alone.[30] From a critical perspective on management of the creative industries, Chris Bilton similarly notes a shift from an emphasis on content to its forms of delivery. Contrasting the current situation to the days in the late 1990s when industry operators believed that "content is king," Bilton writes: "Attention has shifted from the *what* of content to the *how* of delivery, branding, and customer relationships—in other words, towards management."[31] By the same token, just as game software and game platform are codependent, so too the monetization of contents and the success of platforms are codependent. I-mode's success is usually explained in relation to the contents-rich experience it offered. Its lucrative contents market fostered a bubble that in turn produced some of the most important Japanese platforms today: social gaming giants GREE and Mobage, chat app and platform LINE, and the video streaming site Niconico Video all produce their own content, and with the exception of LINE they grew thanks to their position as contents producers for *keitai* mobile phones (in LINE's case it aimed for the smartphone market but used many of the elements found in i-mode). These, in turn, have impacted and are impacted by other Asian platforms, including the Chinese streaming site Bilibili, and Chinese and Korean chat apps WeChat and KakaoTalk. Platforms beget contents that in turn beget platforms.

Complementing this platform-building practice, management writers in Japan also theorized the platform at a very early moment. Within the platform theory that emerges from Japanese management studies in the early 1990s, a platform is both something technological premised on a layered structure (much as software runs on a hardware platform) and, crucially, a mediating mechanism or intermediary between multiple parties, usually in the context of a financial transaction. This latter sense of platform as intermediary is now one of the mainstream denotations of the term within management studies in North America and Europe, within the subfield of microeconomics, and is increasingly prominent within critical platform studies as well. Hence, i-mode does not so much invent the platform as it actualizes or effectuates what was already incipient or projected within management theory, albeit in a manner quite unique, and with major repercussions for the state of mobile media and indeed the internet in toto today—not only in Japan but also, via its channeling into Android and iOS, around much of the world.

The premise of this book is this, then: if we want to know why the term *platform* is so prominent in our time, we must pay attention to the developments of platform theory and practices in Japan. To put it more forcefully: we cannot understand the development of the platform economy until we route the history of the platform through Japan. Yet as this project will make clear, giving an account of platforms from Japan does not mean to ignore the rest of the world. This aspires to be a transnational and comparative account of platform genesis, tracking where and how platforms develop in Japan, the various intersections of U.S. and French discourse that overlap with the Japanese platform project, and the subsequent globalization of the platform via Apple and Google's mobile phone projects. It moves from the national and transnational construction of the i-mode mobile internet platform in Japan, to the global markets forged by iOS and Android, back to the regional development of chat apps in East Asia. This is, as the subtitle says, a story about Japanese platforms in the context of their transformation of the global commercial internet. (With the term *commercial internet,* I refer to the internet as a transactional, commercial space rather than the other military, educational, or community-driven aspects of the internet.) Concretely, this book examines the following aspects of digital culture in turn: *contents* as term in chapter 1, exploring the meaning and development of the term prior to platform discourse; the emergence of platform discourse and platform theory in North America and Japan, focusing on hardware and contents platforms in chapter 2; and turning to the development of the transactional theory of the platform in chapter 3. In chapter 4 I turn to platform building in the case of the Japanese mobile internet systems and i-mode in particular; in chapter 5 we see how the paid-for contents market that i-mode created in turn gave rise to some of Japan's most important platform companies (namely the social game developers and platforms GREE and Mobage, and Niconico Video, the YouTube of Japan), and analyze the unique synthesis of contents and platforms that marks Japan's contemporary platform producers, such as Dwango's Niconico Video. Finally, the conclusion addresses the regional development of chat apps as platforms, looking at Japan's LINE, Korea's KakaoTalk, and China's WeChat in a comparative light, and positioning these apps as the inheritors of the i-mode platform concept. As can be glimpsed from this trajectory, this book shifts among the national, transnational, comparative, global, and regional scales, each of which allows us to tell a different aspect of the platform story. This national-transnational account of the platform routed through Japan's mobile internet experiment draws inspiration from mobile media scholar Gerald Goggin's persuasive approach to the mobile phone as an exemplary media object that requires a global approach.[32]

What begins in Japan with i-mode becomes the agent of global market penetration in the form of iPhone/iOS and Google/Android, which currently have a 99 percent market share for smartphones. This book narrates a view onto the global development of platforms and contents from the particular angle of Japanese platform development, and its associated contents bubble.[33] In recounting the platform story of Japan, this book tells a story that has relevance to anywhere in the world experiencing the rise of the platform-mediated commercial internet, and the attendant transformation of lifeworlds that this brings.

The Japanese Corporate Genesis of the Platform Concept

This book purposefully errs, then, on the side of an overemphasis on Japanese corporations, management writers, economic actors, and platform developers as the loci for platform production, past and present—albeit with a concluding shift to consider the wider East Asia region. This emphasis is a necessary countermeasure to the presentism of tech writing and the corresponding fixation on the United States alone as the site of platform production and platform politics. Of course, the danger here is of replacing the by-now familiar narrative of American platform expansionism with a Japanese one, replacing one imperial power (the United States) with another imperial power (Japan), whose history of imperialist expansion into East Asia remains an open wound in the region to this day. That said, there are reasons for this focus on Japan, and reasons why the economic model of platforms develops in Japan when it does.

The larger context for the development of industrial platform theory and practice in Japan—what we might best describe as the *Japanese corporate genesis of the platform concept*—requires an understanding of how early theorists of platforms were preoccupied by what they saw as a shift from an automobile-centric industrial economy (albeit one already shifting under post-Fordist and Toyotist modes of production) to a postindustrial information economy. The genesis of the platform concept takes place alongside developments in hardware manufacture and insights gleaned from the adoption of technologies. The theoretical and organizational shift to a model of the postindustrial economy had been underway since at least the 1960s and was the subject of frequent white papers and government reports, media theorization, and management theory.[34] Writings on the information society and the information industries start with some fanfare in the 1960s and gain pace during the 1970s. By the 1990s era of platform theory, writers were already primed to regard developments in the information technology sector as integral for guiding Japan's industrial transformations moving forward. This accompanies a general shift in industrial emphasis from hardware to cultural "software" or "contents" that was taking place over the 1980s, due in part to the growing understanding that *contents* ownership was what led to the adoption or failure of technological systems. (Sony's takeaway from losing the VHS–Beta video format wars of the late 1970s and early 1980s was that it needed to acquire and control contents producers, hence its acquisition of Columbia Pictures in 1989; soon after, its hardware rival Matsushita acquired MCA.)

In parallel to the above, the lesson of the 1980s for Japan's hardware makers like Fujitsū was that formerly expensive hardware like computers would be commoditized (i.e., become increasingly a battle to the bottom price-wise, eating into profits) and that the future for computer companies lay not in hardware per se but rather in the Microsoft model of selling software, or cultural contents. A final part to this narrative is the longtime importance of national telecommunications company NTT (formerly Den-den Kōsha), which was privatized in 1985. Its mobile unit, NTT Docomo, was at the forefront of introducing mobile services to Japanese consumers, first as pagers, then mobile phones, and finally the mobile internet. One of its most important and lasting initiatives was the platform genesis of i-mode. Yet even before this, NTT had developed experiments in internet services under the aegis of the "Captain System"—a Japanese interactive videotex system that was the less successful contemporary of France's Minitel. These experiments with platforms led management theorists to hone the vocabulary and theory of the platform. This also requires us to track the longitudinal effects of platform development on the Japanese web since then, as well as mobile media generally, which have made Japan's initiatives into our lived, ambient environment.[35]

The era under consideration was one wherein Japan was briefly seen as the world's future. Looking back at this future formation tells us something important about our present platform economy. As Chalmers Johnson notes about a different era of Japan-watching, this future formation sometimes tells more about transformations predicted in a writer's home country than about futurity per se. Johnson has summarized the discourses around Japan's "economic miracle" years from outside Japan as very often being more about "revealing home country failings in light of Japanese achievements" than aimed at explaining the economic miracle on its own terms.[36] Managerial studies around the Japanese automobile sector repeated this gesture, pointing to the Japanese auto manufacture as a site from which to gain managerial know-how, and to correct the ills of American manufacture. This was epitomized in entities like MIT's International Motor Vehicle Program (IMVP), whose mandate was to bring this Japanese manufacturing know-how to the U.S. context (and conversely, to show Japanese companies how they might interface with American workers as they were setting up factories in the United States). Books that came out of it include James Womack, Daniel Jones, and Daniel Roos's classic 1990 business book examining Toyotist versus Fordist models of production, *The Machine That Changed the World*. This analysis of Japan as an example to be copied is replicated in the literature around the mobile internet project of i-mode as platform in the early twenty-first century—placing the i-mode mobile internet alongside Toyotism and the Japanese economic miracle in terms of global impact. In the first years of the century, i-mode was regarded as the future of mobility, and other countries were urged to follow in Japan's footsteps. As Mark McLelland, Haiqing Yu, and Gerard Goggin recently put it: "The innovation of the *i-mode* system that saw Internet connectivity skyrocket in Japan tends to be what is now remembered, at least in the Anglophone literature, about Japan's Internet history."[37] This techno-Orientalist vision of Japan's mobile internet as the future has had real effects, including the impact on the eventual development of the smartphone in the U.S. context.

There are also institutional reasons for the focus on Japan, foremost among them the relatively national quality of telecommunications companies ("telecoms") at the time of the historical genesis of platforms. That this narrative is located mostly in Japan is a methodological decision that mirrors the nation-centric operation of telecommunications companies—typically national monopolies providing telephone and data communications services. In the mobile sphere on which I focus here, this situation changed by the early 2010s with the rise of iOS and Android as globally dominant mobile platforms, which in turn transformed the role of telecoms from the mediators of commerce to what are often called the "dumb pipes" that simply provide telephone and data services for their subscribers. This marks a shift from a moment when national telecoms giants controlled the mobile space to what is a de facto global regime of standards and shared operating systems—in the form of Android and iOS.

Android and iOS are a duopoly that model themselves on the types of mobile phones popular in Japan since 1999: internet-enabled, function-rich phones, linked together by systems of commerce that allowed for the selling of digital commodities (contents) such as ringtones, mobile games, and so on. The narrative told here, then, is one wherein a certain aspect of what was once regarded as the world's future—Japan's mobile technology and uses—becomes the basis for the globally diffused smartphone-enabled commerce we see today by taking Japan out of the equation.

These are the circumstances under which we revisit the Japanese platform moment and its aftermaths. This book took shape at a moment when, alongside the waning of Japan's i-mode system and mobile internet-enabled "feature phones," BlackBerry OS (another important mobile platform precursor, though predominantly organized around providing email in its early years) and Windows phones also went into terminal decline, to the unique benefit of iOS and Android, which became the globally dominant standards. It was written at a moment that also saw the increasing platform domination of local markets by American giants Apple, Google, Facebook, Amazon, and Netflix, with attendant anxieties around this domination, and questions about what this global dominance of U.S.-origin transnational tech companies means for cultural production throughout the world.

The current global hegemony of American platforms raises the specter of what Dal Yong Jin has termed *platform imperialism*, wherein the overwhelming global dominance of U.S. operating systems and web platforms on the global internet landscape and on the power over the distribution of cultural goods requires a resuscitation of the "cultural imperialism" debates of the 1980s and 1990s.[38] Cultural imperialism is a framework deployed in political–economic theories of globalization, wherein concerned critics note the erasure of cultural differences under the sameness of American culture as it was distributed and consumed around the world (particularly via Hollywood films, American television shows, and fast food culture). Fredric Jameson engages closely with this position, arguing that there "is a fundamental dissymmetry between the United States and every other country in the world, not only third-world countries, but even Japan and those of Western Europe."[39] Through institutions like NAFTA and GATT trade agreements, U.S. cultural goods are made to circulate through the world, provoking anxiety around what Jameson calls "the violence of American cultural imperialism and the penetration of Hollywood film and television," which result in the "destruction of those traditions," cultural and otherwise, in target countries.[40] This understanding is given granular political–economic detail in studies like *Global Hollywood*, which exhaustively examines the economic and institutional conditions for Hollywood's global cultural domination.[41] In the late 1990s, cultural studies and postcolonial theorists pushed back against the cultural imperialism framework, advocating a more nuanced approach to Hollywood and American cultural circulation, acknowledging local difference in interpretation, cracks in U.S. mass cultural hegemony, and multidirectional forms of cultural circulation (the popularity of anime, for instance, or Korean dramas). Yet Jin and others reopen the question of cultural imperialism in the context of the increasing platform domination by Silicon Valley giants. Jin writes:

> The idea of platform imperialism refers to an asymmetrical relationship of interdependence between the West, primarily the U.S., and many developing powers—of course, including transnational corporations. . . . Characterized in part by unequal technological exchanges and therefore capital flows, the current state of platform development implies a technological domination of U.S.-based companies that have greatly influenced the majority of people and countries.[42]

In sum, Jin suggests that cultural imperialism may have been succeeded by platform imperialism, requiring a different focus for political–economic analysis. Our attention should therefore shift from the cultural content of goods to the supposedly neutral platforms that mediate these cultural goods—and their disproportionately U.S. ownership.

Should the intermediation of local cultural production by American giants concern us any more than national telecoms dictating what could and could not be seen and engaged with in previous decades? The platform imperialism thesis suggests that we answer in the affirmative. Maybe this *does* matter in ways that we are just beginning to grasp—such as in the ways Netflix's vertical integration and global penetration promises to reshape what counts as contents, for better or worse, or the ways Facebook's data collection and parsing of users into targetable microsegments has had tangible negative effects on elections and political sentiment globally (to put it mildly). The mere posing of the question of platform imperialism is an important starting point for larger dialogues around the current digital media ecology. Indeed this is an issue I will return to in chapter 5 when addressing the particular forms of interface in Japanese platforms.

On the other hand, I would also signal a note of caution around posing the question of American imperialism, insofar as suggesting the endangerment of Japanese local culture both presumes a local focus (or evacuates the power struggles in the production of such a thing) and tends to erase Japan's own imperialism (cultural or otherwise). Indeed, the very counternarrative of the Japan-origin platforms that this book narrates—while important in counterbalancing an overweening focus on U.S. platforms and telling another history of platform development at its best—at worst threatens to repeat a Japanese imperialist narrative of developments radiating from Japan outward to elsewhere in Asia. Nowhere does the imperial narrative of platforms—and the impact of platform and contents discourse—become more clear than in the merger of Kadokawa and Dwango, which I turn to briefly.

A Contents-Platform Merger

To find a concrete example of the articulation of and corporate-level struggle against Silicon Valley's platform domination, the pitfalls of the platform imperialism critique, and the pervasiveness of the contents-platform discourse, we need look no further than the Kadokawa–Dwango merger of 2014. This was a Japanese reprisal of the AOL–Time Warner merger done almost fifteen years earlier, under a different set of circumstances. On October 1, 2014, Kadokawa, one of Japan's largest publishers and aspiring media conglomerates, completed a merger with Dwango, owner and founder of Niconico, Japan's most culturally significant video platform. Pundits speculated about the reasons for this merger and questioned its financial merit. Kadokawa—the adopted name of this merged company—has since announced voluntary retirements and large-scale restructuring; share prices have dropped, leading some to view the merger as simply an attempt to bury the publisher's debt.[43] Yet this merger is fascinating for the ways the explanations for it articulate a media strategy that reflects, on the one hand, the emergence of a new set of keywords for the current moment, and on the other, changing conceptions of media corporations that now see technology companies such as Amazon and Netflix as the behemoths against which they must position themselves.

Significantly for this book, the transliterated English terms *platform* and *contents* animate justifications for this move toward media conglomeration.

At the May 14, 2014, press conference where the merger was initially announced, Dwango founder and CEO and former Kadokawa chairman, Kawakami Nobuo, explained the terms of the merger in the following manner: "Kadokawa and Dwango have both developed contents and platforms—Kadokawa for the real world, and Dwango for the world of the Net. Having offered both contents and platforms for the net world and the real world, the compatibility of these two companies united as one is extremely good."[44] The merger promised to create what CEO Kadokawa Tsuguhiko effused as "the Platform of the Rising Sun" *(hi no maru no purattofōmu)*—the "rising sun" being a nationalist code word for Japan during the wartime, imperial era of the 1930s to 1940s, during Japan's imperialist incursion into Asia.[45] That this term returns at this moment of time is symbolic insofar as it evokes an oppositional narrative of Japanese versus U.S. platforms (sometimes figured as invading "black ships," recalling the Western countries' forced opening of Japan in the nineteenth century) that ultimately supports a nativist nationalism. It is in this nationalism that we also see most clearly the pitfalls of the platform imperialism hypothesis; even as it draws attention to the political–economic dominance of U.S.-origin platforms, it also authorizes a problematic cultural and economic nationalism that easily falls into line with the Cool Japan propaganda of the government. In doing so it shores up the Cold War paradigm of Japan–U.S. relations and what Naoki Sakai has called the "schema of co-figuration," wherein Japan always defines itself in relation to the West, or the United States.[46]

The Kadokawa–Dwango merger, its background in Japanese platform development, and the terminological explanation of it build on years of transformations in how media work and how media are discussed in Japan; this merger also marks the consolidation of the story told in this book about the rise of Japanese platform theory and its relationship to Japanese contents and platform practice. Kadokawa the publisher and media conglomerate embraced the term *contents* and its strategic media implications earlier in the century; Dwango for its part benefited immensely from one of the earliest implementations of the platform concept in the form of NTT Docomo's i-mode project. As I have suggested, the prominence of the terms *platform* and *contents* within the explanation of the merger must also be situated within a geopolitics of increasing platform domination by North American companies; but in part it was also directed at a series of stakeholders—from shareholders to media corporations to the government—who were themselves enraptured by the familiar terms *contents* and *platforms*. Particular though this merger was, the keywords deployed to articulate the rationale of the merger are symptomatic of their hold over their target audience. They also signal shifts that lead to the rekindling of a dream dreamt in the 1990s in North America and Europe about the coming merger of content and distribution pipelines, one that fueled the failed merger between Time Warner and AOL in 2000. The failed future of Time Warner and AOL was rebooted in the Kadokawa–Dwango merger, this time within the frame of contents and platforms. The Kadokawa–Dwango merger hence reflects the organizational power and causal efficacy of the very terms *contents* and *platforms,* and the managerial discourses and practices that subtend them. This book sets out to interrogate this vocabulary, the better to understand the transformative power of these words along with the associated business strategies and practices embedded within them.

Platformization and Method

A model of imperial expansion or platform enclosure also animates discussion around the transformation or enclosure of the web, which we find in both popular journalistic accounts and critical analyses. "The Web Is Dead," proclaimed *Wired* magazine on the cover of its September 2010 issue. Others, slightly less dramatically, describe the enclosure of the web under the rubric of *platformization*. In his introduction to a special issue on platform politics, Joss Hands writes:

> The Internet is vanishing: as its ubiquity increases, it has also become less and less visible in the production and experiences of network culture. Indeed, many of the operations that used to typify the Internet are now funnelled through so-called "platforms." We do not have a single Internet anymore, but rather a multiplicity of distinct platforms, which . . . are broadly defined as online "cloud"-based software modules that act as portals to diverse kinds of information, with nested applications that aggregate content, often generated by "users" themselves.[47]

In a similar vein, Anne Helmond has written persuasively about the increasing "platformization of the web" attendant to the rise of Facebook and its tentacular APIs that have reconfigured the internet into a trackable and increasingly walled-garden-style space.[48] As new as this phenomenon is, this platformization of the internet itself is also a *return* to an earlier model of the web, one found in AOL on the one hand and the Japanese mobile web on the other. This historical repetition or recurrence requires a careful consideration of what exactly is being repeated—that is, the walled-garden paradigm of AOL and Japan's mobile phones.

Unlike the earlier keyword *network,* which offered a sense of openness, freedom, and rhizomatic extensivity that preempts efforts to control the network, the platform concept is generally applied to the definitive closure of the network, the reigning in of a moment of perceived freedom that the open web was to offer (debates I revisit in chapter 2). For this reason, looking at an earlier moment of platformization provides a critical antidote to the more ahistorical or romantic narratives of decline. It is this need to revisit the Japanese mobile web experience as a formative moment in platformization that requires us to tell a rather different story than that early and prescient one told by the editors and authors of a groundbreaking volume on Japanese mobile phone culture, Mizuko Ito, Daisuke Okabe, and Misa Matsuda's *Personal Portable Pedestrian.* Examining the immense cultural impact of i-mode, and focusing in particular on the lives of its users, *Personal Portable Pedestrian* refuses to adopt the futurist narrative that Japan's then-present form of mobile culture would necessarily become the blueprint for mobile communications around the world—a popular sentiment at the time, expressed in the Japanese press as well as the breathless global media coverage of Japan's mobile technology and culture.[49] As Mizuko Ito writes, *Personal Portable Pedestrian* aimed "toward a certain parochialization and grounding of the Japanese case. The development of keitai [Japanese mobile] uses and cultures is a complex alchemy of technological, social, cultural, economic, and historical factors that make wholesale transplantation difficult."[50]

Now, more than a decade and a half later, we need to revisit i-mode, under a different set of historical circumstances and with a different critical imperative. The global rollout of Android phones and iPhones (while by no means evenly distributed) has expanded aspects of the Japanese mobile phone experience such that it now impacts global mobile culture.[51] Among other things this includes a generalization of the sense that one should pay for digital contents on the internet—something key to the Japanese mobile experience but far less central to other geographies where the PC-centric "free" internet has become the norm. If this is a story about the *platform business* of Japanese phones, it is in part because the legacy of this business model informs digital platforms today. In this sense I follow the lead of Noriko Manabe, who places emphasis on the business models of the i-mode experiment, arguing, "It was the business model and technological infrastructure that made these Japanese innovations viable in the crucial experimental stages."[52] The Japanese mobile phone has also had a lasting impact on the web experience there. It has been a core element to Japanese internet culture in general, as a forerunner of mobile e-commerce, a generative zone from which some of its most prominent game and video companies emerged, and the source for the current vocabularies used to describe it. It also had a transformative effect—via iOS and Android—on the global commercial internet as a whole.

In sum, in light of the platformization of the internet taking place today, it is time to de-parochialize the Japanese mobile internet story, revisiting the crucial role mobile technological developments and their business models played in the language, theory, and practice of Japan's digital business and media ecology. It is this story of the platformization and commercialization of the internet that this book tells, through the angle of Japanese managerial discourse and practice around the term. It is also the (inevitably partial) story of Japan's current media ecology, as it is impacted by the language of contents and platforms.[53]

In looking for models of how to analyze the relation between management and media practice, and the impact of words on worlds, I draw methodological inspiration from works that take the impact of managerial thinking seriously. Two such sources already mentioned are Boltanski and Chiapello's sociology of capitalism in *The New Spirit of Capitalism* and Liu's close analysis in *The Laws of Cool* of managerial discourse, out of which he reads a new image of the subject, and the model of the "cool" workplace within knowledge work in Silicon Valley. Another is

Wendy Chun's careful attention to the intersection of discourses, their ideological presumptions, and computational practices across her work on new media in *Programmed Visions* and *Updating to Remain the Same*. Chun interrogates how language matters, how words and metaphors have effects, demonstrating how one must range across popular and scientific discourses and practices to arrive at an image of new media, or software, or—in the case of this book—the platform. Recent works on film and media theory press us to think more broadly about the sources and sites of media theorization.[54] One of these alternative sites of theorization must certainly be publications from management studies, whose work on platforms I read both symptomatically and media theoretically. Recent work in media industry studies, such as Derek Johnson, Derek Kompare, and Avi Santo's edited collection *Making Media Work*, has led the way in arguing for the need to pay attention to practices of management and managerial discourse.[55]

Media studies of late has kept a close eye on technologies and infrastructures, data and design. It has paid less attention to the economics of platforms, and the business discourses around them. Media industry studies is the exception to this, with work by Jennifer Holt, Alisa Perren, Alexander Zahlten, and others continually pointing to the importance of engaging with industry discourse—and this book positions itself within this subfield.[56] That said, when media studies addresses business, it often does so through the frameworks of labor, political economy, and ethnographies of production. While this book is indebted to the pathways opened by all three of these approaches, it takes a more expansive view of work from economics and management studies—treating it both as ideological formations *and* as a form of theorization worth taking seriously, if for no other reason than that it is productive of platform realities. As such, this book engages with the vocabulary and business strategies of media industries more than the labors, technologies, or ground-level practices, heeding David Nieborg's call to better integrate political-economic and management studies approaches to network businesses.[57]

The Platform Economy ranges widely in order to arrive at a holistic account of the platform that treats managerial theory, platform-building practice, and the impacts of both on contemporary platforms. It offers an account of Japan's mobile internet that traces its impact inside and outside the mobile market, offering a different history of the internet (heeding a crucial call for internationalizing internet studies, to borrow the title and argument of Gerald Goggin and Mark McLelland's 2009 book), and a unique account of platforms.[58] This account embraces what Michelle Cho calls "the existence of multiple internets that are determined by language, regulatory measures, software interfaces, and communications infrastructures," showing the contours of the paid model of the internet as it emerges in Japan.[59] It equally draws on the rich discussions around platforms, from Tarleton Gillespie's and José van Dijck's rich accounts of connective media, to Bogost and Montfort's hardware-centric work on platform studies, to Helmond's work on the platformization of the web, to Jin's political-economic reading of platforms, to Lamarre's work on regional platforms.[60] Present here too is a critical engagement with the extensive publications of Japan's internet business elite, from Dwango CEO Kawakami Nobuo, to i-mode cocreators Natsuno Takeshi and Matsunaga Mari, to Kadokawa CEO Kadokawa Tsuguhiko. These provide not only firsthand accounts of the systems they create but also testaments to the discursive power these words possess, and the effects they have on business thinking and strategy. Finally, this book situates itself within a lineage of critical inquiries into the history of Japanese management and industrial formations, including Simon Partner's *Assembled in Japan* and Chalmers Johnson's *MITI and the Japanese Miracle*.

What emerges out of this is a prehistory of the platform as we know it—a retelling of the emergence of platform theory, an examination of emergent platform practice around Japan's internet-enabled mobile phones, and a tracing out of the impacts of the mobile phone experiment on Japanese and global commercial internet cultures.

Overview

The platform is, then, the central object of engagement of this book. Yet platform theory and discourse cannot be understood without recourse to the discourse around contents—the things that platforms carry and contain. Chapter 1, "Contents Discourse: A Platform Prelude," hence offers a thorough consideration of *contents,* a term that emerges in North America prior to platforms and designates something like the commodity value of a cultural good in a postdigital world. *Contents* is a term that comes into its own at a time when (1) anxieties about the ability to monetize cultural products (from films to books to video games) coincide with (2) the prospects of new, digitally mediated markets and transaction settlement for the products, and (3) the rise of a powerful new delivery platform in the form of i-mode. This chapter examines the rise of contents discourse in the early 1990s in Japan (albeit with some attention to the U.S. context as well), and the manner in which the term becomes attached to cultural goods representative of Japanese cultural cache—namely manga, anime, and video games. In a word, *contents* (as singular plural) designates the commodity value of a cultural good in a digital world. It is a term to which we will return throughout the rest of the book, particularly when seeing the rollout of i-mode. This chapter hence lays critical groundwork for subsequent analysis of its partner term, *platform.*

Chapter 2 shifts to "Platform Typology: From Hardware to Contents," offering a synthetic and typological account of existing work on platforms within media discourse and media studies, both within the Anglophone context as well as in Japan. Modifying a typology proposed by Japanese management writers Negoro Tatsuyuki and Ajiro Satoshi, I suggest we group the variety of usages of the term *platform* into three main types: (1) product-technology platforms; (2) contents platforms; (3) transactional or mediation platforms.[61] This chapter engages the first two types, highlighting the tension between the earlier hardware-centric use of the term and its subsequent metamorphosis whereupon it references social media sites like YouTube, Facebook, or Twitter. For the hardware-centered use of the term I will examine definitions by venture capitalists such as Marc Andreessen and, with a similar attention to the computational, within Ian Bogost and Nick Montfort's Platform Studies MIT Press book series. Within the product-technology platform I will also note its use within the automobile industry, where the platform refers to a common chassis used among various car models. The latter is important not only for expanding beyond the computational but also for understanding the genealogy of platform-related theorization in Japan and in the United States, where this automobile-related definition gained particular traction. As it turns out, automobile manufacture is the heartland of platform research and theory; one strand of platform theory emerges out of an attempt by scholars affiliated with MIT's International Motor Vehicle Program to capture the specificity of Toyotist models of car manufacture. Contents platforms, in turn, designates a stream of objects that we know as social media sites or apps, things like YouTube, Facebook, and Twitter, which have given the term *platform* new meaning and have arguably eclipsed the computational usage in general. Ultimately, however, I will suggest that we need to look to a separate body of work to access the broader sense in which platform is used today, particularly within the media industries: management literature.

Chapter 3, "Transactional Platform Theory: A Japanese Genesis," undertakes an intervention in media studies accounts of the platform by turning to Japanese management literature and American and European work on multisided markets, both of which flesh out a theory of the platform as a mediatory or intermediate function between markets or between third parties. Credit cards, auction sites, and third-party automobile parts suppliers all become examples of this alternate definition of the platform, one that arguably has the greatest influence on industry leaders' invocation of the term today. Further, I demonstrate in this chapter that this literature gets its start in Japanese managerial discourse in 1994, before moving to France and the United States in literature on multisided markets around the year 2000. This chapter explains the importance of both bodies of work and suggests that the early development of this theory in the Japanese managerial context impacted the implementation of the i-mode project in the late 1990s. The chapter further examines how mediation platforms undergird an expansion of management practices into all aspects of digital life—from the mobilization of users for the generation of contents to the management of uses under digital rights management (DRM). By being the mediator or the "middle," the platform operator asserts power over all entities it mediates—whether these are other companies, users, the economy, or society. This chapter hence uncovers a logic of "mediation management" that binds all platforms today.

Chapter 4, "Docomo's i-mode: Formatting the Mobile Internet," shifts our focus from platform theory and its development of mediation management to platform-building practice, offering an account of the development of the world's first successful rollout of the mobile internet. It examines the key players in the i-mode project, as well as its conceptualization as a trans-action platform that would, in turn, become the basis for other businesses (and is a hybrid of all three types of platforms studied in earlier chapters). Of particular interest is the crucial relationship between platform provider (Docomo) and contents providers (those companies that provided services or data products to consumers, from airline reservation booking sites, to banking, to wallpaper downloads and gaming). Through this structure, Docomo created one of the first successful *paid-contents* service environment for the mobile internet, leading many of its contents providers to launch initial public offerings (IPOs) and go public, and ultimately having a great impact on the eventual implementation of platform capitalist enterprises. The resulting "contents bubble"—a moment when low-value contents could be sold at high prices, and wherein the stock market valuation of the contents producers was similarly inflated—reinforced the already entrenched term, leading to an increased sense that there was money to be made in contents in a virtuous circle that propped up all business parties involved. The history of i-mode as the project that formatted the mobile internet experience in Japan and served as a model for both Apple's iPhone and Google's Android is one that needs to be recalled as we have shifted from a web-based experience of the internet to an app-based internet experience that is very much dependent on the paid, commercial model of the internet that i-mode pioneered. (Indeed, the very term *app* likely derives from i-mode's earlier shortening of "application" to "appli.")

Chapter 5, "Platforms after i-mode: Dwango's Niconico Video," returns to the Kadokawa–Dwango merger, which I touched on above and whose present activities are in large part indebted to the i-mode experience. This chapter presents a wider examination of the evolution of contents pro-viders into platforms builders in their own right, seeing this occurrence as part of a wider *i-mode effect*. Here we move from the story of i-mode to the manner in which i-mode functioned as a basin for activity on or around it, including being the birthing ground for some of the major social media applications that sprung from its platform. DeNA, GREE, and Dwango are the three most renowned of these. DeNA was founded by one of the McKinsey & Company consultants who worked on the i-mode project (foregrounding the role of global knowledge purveyors in fostering i-mode itself), and later founded an auction site and a social gaming site. While GREE has declined in importance, DeNA and its subsidiary Cygames remain some of the most important social gaming publishers in Japan today. Subsequently, the chapter turns to examine arguments made about the specificity of platforms in Japan, paying particular attention to Dwango's Niconico Video, Japan's premier video-sharing service. Through an analysis of the unique interface of the Niconico streaming platform, this chapter suggests that one of the characteristics of many Japanese platforms is that they both combine the intermediary, hosting function of the platform as well as create or mediate the contents produced. There is a convergence between contents and platforms that marks Japanese contents in particular. Examining theories of the interface and comments culture within digital media studies, this chapter examines the intertwining of contents and platforms in Japan, and concludes with a consideration of the concept of platform capitalism itself.

The conclusion, "The Platformization of Regional Chat Apps," examines the continuing and unfolding of the i-mode effect as platform model, this time in the context of East Asian chat apps first released between 2009 and 2011: Korea's KakaoTalk, Korea/Japan's LINE, and China's WeChat. We increasingly read about the platformization of chat apps with articles focusing on Facebook Messenger and WhatsApp. Yet these Asian chat apps were the pioneers of this transformation of the humble chat app into a metaplatform that offers all the functionality of the smartphone within the app itself—from calling and texting, to sending money, making payments, reading news, and ordering cabs. This chapter examines these apps both in light of the legacy of the i-mode platform—framing them as reboots of the i-mode concept within a chat app environment—and through the lens of a modified conception of media regionalism. That these are apps distributed on iOS or Android also marks a transformative shift in the circulation and pervasiveness of the platform concept—from i-mode to iOS/Android.

To any who have asked about the meaning of the keywords examined here—*platform* and *contents*—this book seeks to provide an account that ranges across the very heterogeneous sites of these terms' impact, and narrates at least one version of the emergence of the platform economy. It is an answer that avoids the geographical bias of the habitually Silicon Valley–centric writings on the platform (whether critical or celebratory), and the presentism of these writings. It charts the emergence of the term as platform theory, and the practice of building platforms in the Japanese context, offering a historical trajectory of the rise of the platform economy and a critical assessment of its impacts on the management of contemporary life. History—even near history like narrated in this book—provides us with perspective on the world, enabling an informed critique of the current era of platform domination, and a clearer sense of how it is we got here. Displacing network, platform is the new communicative and managerial logic of our time; it is also the dominant commercial logic of the internet. The Japanese platform story is a key moment in how we got to this point, how we acquired the habit of paying for contents on the internet, and how the management of things and words, contents and people, by mediation platforms, developed. It also thereby offers a unique vantage point from which to grasp our platform present. The platform as mediatory device or mechanism rewrites prior relations between companies and consumers, such that economy and society are now governed by platformic relations, which is to say: defined by the management of social, economic, and cultural relations by means of platform mediation. As such, the focus of *The Platform Economy* is less on particular platform entities and more on the managerial logics embedded in particular platform entities (like i-mode), which in turn extend beyond them through their discursive and medial effects. These logics increasingly penetrate a social field now reformatted as the platform society, and animate linguistic and industrial organizations reformatted as the platform economy. This expansionary logic of the platform and its management of companies, workers, products, users, and everything in-between is the true object of this book, and is the referent of that still ambiguous term, *platform capitalism.*

CHAPTER ONE

Contents Discourse

A Platform Prelude

C ontents has overtaken the world. Not content, mind you, but contents
(kontentsu) in the ambiguous plural form that is uniformly used in
Japanese transliteration, ingesting previous media and informational objects as diverse as movies and radio and news, as well as their containers,
like newspapers, DVDs, magazines, and books. The term has a symbiotic
relationship with the keyword of this book, *platform,* but the discourse
of contents emerges earlier, and stabilizes before the term *platform* really
comes into the mainstream. To anticipate my conclusion: *contents* in Japan
is hence never configured as a singular content—a manner of thinking that
Bharat Anand in a recent business book calls "the content trap."[1] Rather,
contents is always figured as plural, and the term invokes the synergies
that mark media franchises as well as the network effects that mark platforms. Moreover, the very possibility of selling contents—that is to say,
of making data or cultural products sellable—hinges upon the bounding
properties of platforms, the operational closure of "walled gardens" that
prevent the free flow of digital files and allow for the monetization of digital content.[2] A strong platform will in turn create the technological conditions and market conditions for the packaging and selling of cultural
goods as contents. Platforms hence create the conditions that allow for
contents to appear as such. And yet platforms are only useful as a means
of packaging something that we call *contents.* As the later accounts of platform building via mobile phones will demonstrate, contents were always

key to function of the platform. Contents were also a lure to draw people to a platform in the first place. We hence start our account of platforms here, with its symbiotic pair term, its indispensable lure: contents.[3]

The incredible diffusion of the term *contents* is particularly visible in Japan, not the least because the word is marked as a loanword, rendered in the katakana script. Over the first decade of the twenty-first century, Japanese writings on contemporary media culture and media industries expanded exponentially. From academic studies, to government reports, to corporate white paper *(hakusho)* research documents, and everything in between, the variety and range of discourse on media in the medium of print is astounding.[4] Equally astounding is the pervasiveness of a peculiar word that runs through most of this diverse array of work. Open any Japanese publication on anime, manga, film, video games, or media in general written since the year 2000 and you will come across a common term used to describe these and other media forms: contents *(kontentsu)*. Deployed with beguiling self-evidence in a range of critical and business texts, contents has infiltrated everyday language—and most discussions of media in the early twenty-first century, especially—in a rather remarkable fashion. Indeed, it functions as a portal through which anyone who dares to speak about media must pass. For all its ubiquity, though, it has received too little critical interrogation. There has been a wave of very strong work on Japanese media culture from a critical vantage point over the last decade— yet few texts interrogate this now ubiquitous keyword, *contents,* with the notable exception of Kuhkee Choo's analysis of Cool Japan, and work by Thomas Lamarre.[5] The time has come to undertake a critical reflection on the rise of the term *contents,* particularly since such a reflection is a precondition for understanding the term *platform* and the intricate relations between contents and platforms mapped in later chapters. Indeed, in order to understand what the platform is and does, one must first work through the prehistory of the platform inasmuch as it is embedded in the rise of the term *contents.*

Sometimes transliterated back into English as content in the singular, the Japanese term notably uses the English transliteration in the plural form: contents rather than content *(kontentsu* rather than *kontento).*[6] The term *contents* has become a discursive locus of governmental policy, private ventures, and public discussion in Japan, South Korea, and China—not to mention the United States and Canada—in distinct ways. In fact, we might say *contents* is one of the most important terms for media production in East Asia today, second only to its sister term, *platform*. Indeed, the significance of the term for regional media production is matched only by its ambiguity in meaning. It appeared suddenly and quickly become the locus of business writing, government policies—from Cool Japan to the Korean Wave—and even academic writing on media in Asia. But what does this term mean, and what effects is it having in these multiple national contexts? What is contents doing?

It is most helpful to look at this from the site of Japan—and, albeit in brief, Korea as well—precisely because the terms *kontentsu* (コンテンツ in Japanese) and *k'ont'ench'ŭ* (콘텐츠 in Korean) are not native but rather loanwords, creating a certain mystique around the terms, and also begging for explanation. It is in the explanation that we find some of the peculiarities of the term, which will help Anglophone readers also grasp what is at stake in its rise and proliferation in other contexts.

As we will see over the course of this book, contents arises contemporaneously with platform as its partner term—and, from the vantage point of the present, neither can be analyzed in isolation from the other. That said, the two terms do in fact emerge independently. In its early days the terms most associated with contents are software and multimedia—computational terms, yes, but not directly linked to the terminology and problematics of the platform. Contents and platforms only converge around the time of the i-mode experiment. It is the success of the i-mode mobile internet and its competing systems that creates the largest market for digitally delivered contents, ultimately binding contents to platforms, and effectively launching the contents bubble of the early twenty-first century. This bubble is both discursive and financial—one feeding into the other. Chapter 5 will examine the bubble that arose in the wake of the monetization of contents with the mobile internet systems launched by mobile carriers in 1999.

This chapter will examine the effects and historical conditions of contents discourse, paying close attention to how the particularities of Japan's media historical conditions impacted the adoption of what seems to be merely another instance of global transliterated English. The story of contents and its rise to prominence since the mid-1990s is firmly anchored in Japan's media-cultural conditions. Out of the unique constellation of historical factors that guide the use of the term *contents,* three founding conditions or moments are the focus here: (1) the emergence of *kontentsu* as a term for media within the context of the digital shift, including factors such as the attendant challenge posed to the very conception of a medium in the mid-1990s, the way the term takes over from an earlier such term, *sofuto,* or software, and the growing need for an alternative revenue stream to computer hardware; (2) the use of the term *contents* to refer to animation, manga, films, video games, and other media mix forms in the early twenty-first century; and (3) the transformation of marketing discourse from a concern with narrative to a concern with contents from the late 1980s to around 2005, which is at once a symptom of the rise of contents discourse, as well as a site from which to grasp the semantic parameters of the term. By examining these three moments together, this chapter will provide a basic understanding of the semantic range and discursive function of the odd term *contents* as it operates within the cultural industries of contemporary Japan. This understanding will in turn help us down the road as we grapple with the monetization of contents in the wake of the groundbreaking mobile internet platforms of Japan's telecommunications giants NTT Docomo (with i-mode), J-Phone (with J-Sky), and au (with EZweb).[7] Contents as monetizable packages of intellectual property allow for the birth of platforms as containment mechanisms and payment structures through which data or information becomes sellable as contents.

Genealogy of Contents

The terms *content* in English and its pluralized and transliterated form *kontentsu* in Japanese arise pretty much simultaneously in North America and Japan in the mid-1990s, and the presence of this term is felt strongly in other Asian countries, South Korea in particular. The emergence of the term *contents* designates a global transformation, a digital era of media consumption wherein the material consistency of an informational or audiovisual commodity has become unhinged from a particular physical medium and mobile in a way it was not before. In a word: this is the phenomenon of digital dematerialization (or, more accurately, since the digital is also very material, the delinking of a given medium from a particular medial substrate).

To speak in concrete terms, using the web platform of Niconico, one may purchase an electronic version of a popular manga, as well as a television anime based on the same manga, and view these contents on any number of devices, whether laptop, television, tablet, or smartphone. The digital packaging, or form of each content (a book volume or a TV episode), remains discrete while viewable across different devices and platforms as digitally encoded data. This is the historically novel phenomenon of what one early writer described as "so-called media-neutral 'content.'"[8] That is, a particular media form and content (an episode of a drama, say), presents itself as medium independent, or *medium agnostic,* reliant on the "metamedium" of the computer.[9]

Contents Discourse in North America

As Paul Frosh puts it in his analysis of the North American "visual content industry" and Getty Images in particular, the term *content* (here in the singular) emerges as a response to "the dismantling of the technical boundaries between previously distinct media (photography, painting and drawing, film, video) and their convergence and mutual convertibility."[10] The term comes to the fore as a response to the need for a new vocabulary that can describe media objects in a medium-agnostic manner, and for an era where the medium-message package is no longer anchored to a single material substrate (such as the film strip of film), or distinct technical apparatuses of production (the 35mm film camera) and sites of exhibition (the film theater). Content and its somewhat later companion term *platform* arise at a moment when a film or a TV show or even a book can be viewed on any number of now ubiquitous computational devices; where a newspaper could be read on paper, in electronic ink, or on a tablet computer—a point when a specific media text is converted into "an abstract universal, 'content'" viewed on screen-based computation devices.[11] Hence, Alan Liu will describe the separation of content from materiality as one of the fundamental characteristics of what he calls "discourse network 2000" (referencing here Friedrich Kittler's analyses of discourse networks 1800 and 1900): "These cardinal needs of transformability, autonomous mobility, and automation resolve at a more general level into what may be identified as the governing ideology of discourse network 2000: *the separation of content from material instantiation or formal presentation.*"[12] This separation had significant implications for the media industries that traded in contents.

In the 1990s, the emphasis in North America was on the production of media content under the slogan "content is king." The term *content* is visible in newspapers and magazines, and business leaders talk about the importance of *content providers.* There is a remarkable increase in the use of the term *content business* and the standalone, singular form *content* in the English-speaking world between the years 1993 and 1995. By 1995 the *Wall Street Journal, The Economist,* trade publications, and even the Organisation for Economic Co-operation and Development (OECD) reports begin using the term with some regularity. In 1994 the terms are predominantly used within quotes from members of the computer world or telecoms industries, and there as "multimedia-content business" or "information-content business"; but by the following year the terms appear to have shed their scare quotes, with journalists discussing the latest trends in the "content business."[13] A 1995 article on the much-touted "information superhighway" that surveys existing business periodicals' account for the term explains, "The high capacity network now in the planning stage will deliver digitized voice, music and full-motion video, including entire movies, as well as files and programs. As a result, businessmen and reporters now talk about 'content' rather than 'information' or 'data.'"[14] In a representative moment, former Microsoft CEO Bill Gates famously titles a 1996 speech "Content Is King" and says: "Content is where I expect much of the real money will be made on the Internet, just as it was in broadcasting."[15] No longer just data or information, film and video and text as content would now fill the internet tubes.

In 1998 William Safire, the late great chronicler of language trends for the *New York Times,* acknowledged a shift in the nature of his employment: "I no longer file copy, or even transmit data; ever at the cutting edge of the pointiest cusp, I *provide content.* If any word in the English language is hot, buzzworthy and finger-snappingly with-it, surpassing even *millennium* in both general discourse and insiderese, that word is *content.*"[16] With

Gates as the representative figure, what we find in early uses of the term *content* is that it comes out of either telecommunication or computing contexts, referencing a kind of generic media as that which is transmissible. Only later—by the late 1990s, if we take Safire to be an early adopter—the term comes into circulation within the culture industries.

Contents Discourse in Japan

This origin of the term within the computer industries impacted the nearly simultaneous take-up of the term in the Japanese context. Content is first and foremost a term used by technologists to talk about cultural goods, *insofar as they are sent through internet infrastructure as data.* Content industries are hereafter redescribed as information providers.[17]

Yet there are two immediately apparent differences between the use of the term *content* in the United States and the term *kontentsu* in Japan. First, while both terms began their life as designating something like "information," or that which could be digitized (pointing to what is undeniably a global media transformation), the term *kontentsu* is quickly given a local inflection.[18] While emerging from the realm of computer hardware firms—not unlike in the United States—in Japan the term *contents* is in a short order associated with particular kinds of content: anime, manga, light novels (a form of young adult fiction that has anime-style characters on the cover of the novels), and games (with film and music on the periphery), which are simultaneously the locus of "media mix" practice, whereby media are conceived and deployed in franchises. They are also the object of governmental promotion in the form of Cool Japan policies.[19] Contents shifts from meaning digital information to becoming a catchword for those media or franchises that were most capable of export—both to increase revenues and to support a "Cool Japan" that had newly discovered the importance of culturally mediated soft power via the early twenty-first-century slogan "Gross National Cool."[20] Moreover, industry players soon take up the term in the context of manga, anime, games, and other forms associated with the phenomenon of the media mix—the Japanese popular and industry term for the transmedia development of particular franchises—a phenomenon widely known in North America by Henry Jenkins's term *convergence culture*.[21] As a result, the term *contents* becomes, by the end of the first decade of the twenty-first century, a catchphrase for government officials, industry actors, and theorists alike.

Contents is perhaps *the* most important keyword designating the digital shift. Its very bagginess and absence of clear definition makes it valuable for understanding how this shift is read, understood, and deployed in Japan. Indeed, we will see that this close association of contents discourse with the manga-anime-game industries is one of the particularities of the term's use in Japan, and one of the reasons for the need to pay close attention to the variations of media keywords within a given geocultural milieu.

A second major difference in the use of the term *content* between Japan and North America relates to its circulation. The term *kontentsu* caught on in both public discourse and the vernacular to a much greater degree than *content* did in North America or the English-speaking world. Or, perhaps more accurately, the term has enjoyed a longevity in the Japanese vernacular and industry discourse since the beginning of the twenty-first century, while the buzz of the word in North America had died down somewhat. In Japan, *contents* has become not only a term used in business publications or web content production manuals; it went mainstream, featured in the titles of academic and popular journalistic work on anime, manga, games, and surrounding media in a manner unseen in the Anglophone world.

Three examples come to mind, even if thousands more vie for attention. The first is the well-known subcultural critic Azuma Hiroki's edited collection *The Thought of Contents: Manga, Anime, and Light Novels* (2007), one of the earlier entries of the term into academic discourse.[22] The second is manga critic Nakano Haruyuki's 2009 publication *Theory of Manga Evolution: The Contents Business Is Born from Manga*, which represents a useful industry analysis of manga fully anchored around the by then dominant term *contents*.[23] And the third is the publication of a book on the where and how of technology, contents, and Cool Japan coauthored by Kadokawa Tsuguhiko, the chairman of media mix giant Kadokawa: *The Cloud Era and the "Cool Revolution."*[24] In fact, Kadokawa's book is an excellent example of the embrace of contents discourse by industry players, as well as a site from which to understand the media and formal effects of discourse; in June 2013 Kadokawa Group Holdings underwent a complete makeover in an effort to become, as company President Satō Tatsuo put it in his January 2013 address to the entire company, "a business group that can develop strong contents globally."[25] The company changed even more radically in 2014 when it merged with Dwango, a games and digital media producer best known for its Niconico Video streaming site—the YouTube of Japan, which I discuss in depth in chapter 5.

Azuma's collection, Nakano's book, Kadokawa's volume, and indeed the very merger of Kadokawa and Dwango all connect the "contents business" with the media of anime, manga, light novels, and games. One of the specificities of Japanese contents discourse is this close association between the particular media forms targeted by the term *contents* and the mainstream practice of the media mix, particularly in its orientation around anime-manga-novel-game franchises. Indeed, many writers argue that the use of the plural form *contents* is unique to Japan, suggesting that the descriptor of "the contents business" was not imported from abroad but originated in Japan.[26] It is to this particular development of the term that I will now turn. A second specificity, to which I turn in later chapters, is the financial and discursive impact of i-mode and the mobile internet on the creation of the "contents bubble."

Contents Is the Keyword for 1995

The term *contents* is first used with some regularity in Japan in the mid-1990s, in reference to the digital shift and the informatization of previously distinct media forms. During the 1980s, one comes across the term *contents* solely in reference to the automobile industry, around negotiations about the quantity of an automobile made in Japan or the United States, indexing a competitive moment within the automobile industry. The first use of the term *contents* in the contemporary sense—generally glossed as *informational content*—comes in 1993, quickly attaining keyword status by 1995. The push toward *contents*—as keyword and strategy—appears to be of information technology company Fujitsu's doing.

One of the first books with the term *kontentsu* in its title is actually a translation of Robert E. Horn's *Mapping Hypertext: The Analysis, Organization, and Display of Knowledge for the Next Generation of On-Line Text and Graphics,* originally published in English in 1989. The title of the 1995 Japanese translation is of interest for the way it surreptitiously inserts the

newly popular term *contents* in its title: *Studies of Hypertextual Informa-*
tion Organization: Recommendations on the Creation of Structural Con-
tents.[27] This book is published by Nikkei BP, a business book publishing
house that put the term firmly within the sphere of corporate strategy.
Presuming that readers would find the term *structural contents* somewhat
bewildering, Nikkei BP's website entry on the book helpfully defines the
presumably unfamiliar term *contents*: "The secrets of the production of
contents *[kontentsu]* (informational content *[jōhō no naiyō,* 情報の内容]),
understood through visuals."[28] Here, contents is equated with informational content, or simply information. As we will see shortly, *content* and
information became an increasingly typical conflation as the term came
into wider use.

Notable here is also the function of the parenthetical gloss on this
keyword. The glosses on this loanword used in an unconventional sense
that appear redundant to an English reader—"contents (informational
content)"—in Japanese call attention to the strange or foreign quality of
the loanword, its semantic malleability, and the need for some explanation. Contrary to the general cultural figuration of parenthesis as a brack-
eting of the playful, unimportant, or marginal away from the important—
a figuration that Jeff Scheible notes and complicates in his book *Digital*
Shift—here the parenthesis is precisely the definitional moment that con-
figures this new term, or inserts a hitherto existing term into a brand new
context.[29] Parenthetical glosses are a definitional moment, a space to provide a key to reading a new word, with novel semantic parameters. They
call attention to the peculiarity of the word; at the same time, they try to
nail down its meaning and introduce it into the daily vocabulary. Their
existence speaks to the novelty of the word, and their eventual disappearance to the word's habituation within everyday language. In 1995 the gloss
was ever-present.

Shimizu Kinnichi's *Fujitsū's Multimedia Business: Business Strategies from*
Infrastructure to Contents, also published in 1995, gives the term *kontentsu*
a related if slightly different spin, associating the term with the media content, or *software* (and its abbreviation as "soft"), rather than the hardware
infrastructure side of the company's multimedia business.[30] It is no coincidence that one of the earliest publications with *contents* in the title is a

book on Fujitsū. Nor is it unusual that all the above books were published in 1995. The driving force behind the introduction of *contents* to Japan and to its proliferation as a keyword appears to be Fujitsū, one of the largest players at this time within Japan's information technology sector.

In 1993, in the earliest usage in the contemporary sense, Fujitsū CEO Sekizawa Tadashi's speech on the occasion of the financial newspaper *Nikkei Shinbun*'s twentieth anniversary emphasizes a shift from a focus on hardware to a focus on multimedia, which he defines as the product of the "convergence of information processing and transmission technology."[31] Suiting the celebratory occasion in anticipation of the massive changes the publishing industries would face in the future, Sekizawa also reflects on the challenges of multimedia.

> The infrastructure for the new society is multimedia itself.
>
> I've talked about the wonderful society that we'll arrive at with multimedia, but before it can come, several social problems must be resolved. The first is how to price information. . . . The second is the problem of [intellectual property] rights around electronic data. The expansion in the circulation of data means that we will need rules about to whom that data belongs. Additionally, there is the [third] issue of the circulation of contents (informational content [情報の内容]). . . . Without solving these three issues it will be difficult to popularize multimedia.[32]

Contents as a keyword only comes up in passing, under the wider umbrella of the multimedia society. That being said, Fujitsū was by then pouring resources into multimedia and contents, establishing a Promotion of Contents Business Division in June 1994 and subsequently creating a contents production company in December 1994.[33]

By 1995 *kontentsu* was picked as the first keyword of the year for the "Keywords for Understanding the Information Communication Industries of 1995" article series by the very same Nikkei newspaper, featuring the Fujitsū CEO yet again in its write-up on the topic. There, the unfamiliar word is given a formulation by now familiar to us: "From films, to broadcast television shows, CD-ROMs . . . and including the content *[naiyō]* of soft *[sofuto,* an abbreviation of software], [the word *contents*] in the widest sense refers to the substance *[nakami]* of soft."[34] With the new distribution system of the commercial internet right around the corner—launched to the Japanese public in 1994—the seemingly staid computer hardware company Fujitsū saw a chance to diversify from hardware production to the racier realm of content production. It mobilized this unfamiliar word to create a new business division, and thereby moved into a domain it had until then been detached from: the production of cultural goods. In so doing it played a key role in pushing the term into prominence not only in Japan but likely in South Korea as well.[35] Shimizu Shin'ichi in *Fujitsū's Multimedia Business* quotes the representative of the Promotion of Contents Business Division, Ochiai Takashi, explaining this shift: "Fujitsū has many business divisions, but they inevitably treat hardware as the basis for business, and involve the deployment of this hardware as a commodity. However, the Contents Division doesn't have any hardware. It's completely and purely about selling soft alone. It's Fujitsū's first foray into cultural production."[36] In a symptom of the shift toward the importance of cultural production and the cultural industries, hardware maker Fujitsū dips its feet into the creation of cultural contents. The provenance of the word *contents*—a hardware manufacturer diversifying into cultural software—will continue to inform the use of the term *contents,* and in particular its function as a linguistic packaging for cultural products in the era of digital convergence.

Indeed, *contents* as a term gestured at a state of affairs similar to what in Anglo-American contexts was called media convergence—the supposed convergence and distribution of multiple distinct media forms into one information channel, and one "black box" from which games, film, television, and music would all be accessible (something realized to a degree but with the caveat that most people possess many such proprietary black boxes—Apple TV, Chromecast, PlayStation, etc.). To glimpse this, it is worth quoting again from Nikkei, this time as it offered a definition of contents in its April 2, 1995, edition of "Word of the Day" ("Kyō no kotoba").

> Contents: Informational substance, content. Can include moving images *[eizō]*, music, news, games and so on more widely, however the representative contents of the multimedia age are often said to be game software

and home shopping as transmitted over cable TV (wired television) and computer communications. CD-ROM software is often offered as an example of multimedia contents.[37]

While its influence in the propagation of the word is clear from the above definition, Fujitsū's forays into content production seem to have borne little fruit. Still, even if Fujitsū is not known today as a major player in the content industries, it undoubtedly did put the word *contents* onto the Japanese information industry map.

Indeed, it is fascinating to look back on the terms highlighted in Nikkei's "Keywords for Understanding the Information Communication Industries of 1995": Contents, Advertising Performance, Windows NT, PHS (personal handy-phone system) business, Home Printers, DOS/V, PC 98 (PC 9800, by NEC), Power PC.[38] Of these, the only truly enduring keyword is *contents*.

Software, Information, Multimedia, Contents

So far, so familiar; this use of contents sounds much like the roughly contemporaneous popularization of the term by hardware and software players in the United States, except something is different in this iteration— and we can begin to understand this difference by homing in on the odd word *sofuto* that preceded *contents*. It is no coincidence that the third and earliest book to be published with *kontentsu* in its title in 1995—Noguchi Hisashi's *Kontentsu bijinesu: Media sofuto wo sagase* [Contents business: Hunt for media software!][39]—also includes the term *soft*. The contiguity between *soft* and *contents* suggested in the title points to another important lineage for the latter term.

Software or its abbreviation *sofuto* was often used as a shorthand for audiovisual information contained in a particular media commodity during the 1980s and 1990s—rather than referring exclusively to applications like Microsoft Word. By the late 1980s and early 1990s, *sofuto* could more widely refer to narrative or other *transferable forms of intellectual property within the media mix*.[40] This usage itself comes from a more concrete, earlier usage in relation to video software—or VHS cassette tapes—that was a burgeoning market in early 1980s Japan. If the video player was the "hard" *(hādo)*, the video cassettes as well as their audiovisual content were the soft. Another essential medium around which the term *soft* is used is video games; a video-game cartridge is called "game soft." Here "soft" is both the cassette as a physical format and the particular game itself. By the late 1980s, soft is used to refer to what we would now call game content. As Thomas Lamarre writes, "Games became 'game software' *[gēmu sofuto]*, and animation on cassette, laser disk, or DVD, became 'video software' or 'anime software' *[anime sofuto]*."[41] A term originally designating the container also came to designate its content.

We can turn back to Kadokawa Books, a book publisher that as of the mid-1970s was becoming a media company, for a concrete example of this use of *sofuto*.[42] In an internal annual report from 1984, Kadokawa Tsuguhiko, brother of then-CEO Kadokawa Haruki, announces that he was preparing the company to keep pace with what he called the "Era of Soft" *(Sofuto no jidai)*, establishing the groundwork for the Media Mix Lab division of the company at that time.[43] In 1990 the Media Mix Lab was expanded and renamed the AV Department—with AV being another stand-in for video. Both the Media Mix Lab and the AV Department were key sites where on-video animation works were produced. These works were based on intellectual properties developed at Kadokawa—comics and novel serializations in particular. In 1994 the AV Department was folded into the "Soft Business Division," incarnating the ambition to pioneer the era of soft.[44]

Hence the term *sofuto* quickly expanded from a video- and computer-related term to something that encompassed all forms of narrative-based intellectual property. While not all games are narrative in form, the games Kadokawa produced certainly were, coming at the height of the explosion in fantasy games and associated media, and inevitably paired with manga tie-ins. In 1997 Kadokawa Tsuguhiko announced that hereafter the company would aim to become a "mega software publisher"—a bold claim for what was formerly a book publisher, and one that could only make sense if we understand software to mean "media" or "intellectual property." By 2004 the company's self-description shifted to incorporate the term *contents* at the core: it was to become a "mega contents provider" *(mega kontentsu purobaidā)* and a "comprehensive media company" *(sōgō media kigyō)*.[45]

Even before its merger with Dwango, Kadokawa had become one of the foremost companies associated with the subcultural media of manga, anime, light novels, and games.

The propensity for the semantic slide from *software/sofuto* as a term referring to a category of media packaging or format (the video cassette or game cartridge) to anime or video games as *kontentsu* is already palpable from the above example. Another clear link between the terms *soft* and *contents* can be found in the *Multimedia White Paper (Maruchimedia hakusho),* first established in 1993 and published yearly until 2000, when its name changed to the *Digital Contents White Paper (Dejitaru kontentsu hakusho).* When first established in 1993, the editorial collective of the white paper called itself the Multimedia Soft Promotion Association. In 1996 it changed its name to the Multimedia Contents Promotion Association.[46] In 2001 the group merged with the Commerce and Information Policy Bureau of Japan's almighty Ministry of Economy, Trade and Industry (METI), at which point the editorial team assumed the title Digital Contents Association. The Cool Japan policy initiative, so integral to Japan's contents policies including the promotion of marketable contents from anime to manga to character goods and maid cafés, is also housed in this very same Commerce and Information Policy Bureau within METI. There is thus a clear transition from "soft" as the organizing principle of media industry and industrial policy to "contents" beginning in the mid-1990s and consolidated in the early twenty-first century. This semantic slide from *soft* to contents is in large part what informs the specific valences of *contents* as a term, referring as it does to particular kinds of media in the Japanese context, and within the Cool Japan initiative in particular.

If software and "soft" is one of the transitional nodes to the term *contents,* "information" is another. One site where we find this link between information and contents articulated most clearly is in the computer industry annual white paper, *Jōhōka hakusho,* or the *Informatization White Paper.* This particular white paper was published between the years 1987 and 2009, by the Japanese Association for the Development of Information Processing (Nihon jōhō shori kaihatsu kyōkai). This was the newer incarnation of an earlier annual white paper, *Computer White Paper (Konpyūta hakusho).* The *Informatization White Paper* is a useful site for considering the transition to contents, particularly for the way that the rise of the term in this white paper is not simply a story of displacement from software but the introduction of contents as a third term between hardware and software. The emergence of contents, therefore, indicates the rise of a novel media condition.

Examining this white paper from the early 1990s onward, there are a number of terms at the nexus of which contents arises: multimedia, services, and software. Notable in the use of multimedia is that it is not a replacement for media per se but a term that designates the capability of hardware to process multiple forms of media and allow users to access them on a single platform. Let us take a look at the definition of multimedia that the *Jōhōka hakusho 1993* offers.

In recent years, we have seen the remarkable development of high-performance, miniaturized and increasingly inexpensive information processing devices. For example, we have seen devices that can read CD-ROMs, which make possible the storing of large amounts of information as external memory; at the level of the PC's operating system, it has become possible to handle moving image and sound. Not limited to text and sound as we were in the past, computers for personal use can now process multimedia information such as still images, moving images, and drawings. This transformation is what is known as multimediazation.[47]

Multimedia here designates the capacity of computers and their users to process and replay media such as moving images and drawings that were previously beyond the capabilities of the computer. A similar description of multimedia is included in the 1994 edition of the white paper, with the added preface that "in recent years, with the increasing cost performance of CPU and memory devices and so on, the aspect of personal computers as *multimedia platforms* has become more pronounced."[48] Of course the deployment of the term *multimedia* was not new to this time, as the following year's report remarks. Indeed, the central government's emphasis on multimedia, the report writes, "recalls the new media boom of the mid-1980s."[49] If this recalled a certain fervor for multimedia from a decade earlier, by the 1996 report this excitement seems to have all but disappeared.

The section heading dedicated to "Multimedia Personal Computers" in earlier years was, as of 1996, replaced by the heading titled "The Outlook on Informatization for Individuals and Their Lives."[50]

The term *contents* appears suddenly and prominently in the 1997 edition of the white paper, as if they could stave off the tide no longer (it bears noting that the 1997 edition is published in June 1997, so around two and a half years after the Nikkei newspaper had declared *contents* the keyword of 1995). The term is allocated its own chapter within the second part of the report, on "The Information Services Industries," titled "Currents in the Contents Business." One of the first mentions of the term, earlier in the volume, reads: "Even as hardware makers have heightened their focus on the software *[sofuto]* and service business, the newly influential enterprise of the contents business *[kontentsu bijinesu]* is emerging. Regardless of the size of the revenue stream of these enterprises, hardware makers have established contents businesses as subsidiaries."[51] Here we find a certain ambiguity about whether the contents business is an equivalent of the software business or whether the contents business is a third realm distinct from both hardware and software in the traditional sense (it seems to be the latter). It is this thirdness of contents as a distinct enterprise that makes the *Jōhōka hakusho*'s first steps into discussing the contents business so interesting.

Moreover, the opening lines of the chapter dedicated to discussing the contents business immediately point to the novelty and the prevalence of the term.

It has not been very long since the word "contents" started being used with regularity, and one now comes across the term on the pages of newspapers frequently. It is used with a fairly wide range of meanings, from the narrow sense of "software such as games, music and so on" to the rather broad sense of "material composed of sounds and images." However, why is it *now* that contents have gained such attention?[52]

Not wasting a moment, the writers offer an answer to their question, one that works to disentangle software from contents and from multimedia.

Given the seeming substitutability of contents for software (as I argued above), it is worth quoting this response at length.

> Contents have gained such attention now precisely because we have arrived at a time when the shape of multimedia has become concretized at the level of commodities and services. . . . Software refers to information that has taken relatively organized form. Contents, on the other hand, has a rather wide sense, meaning anything from as small as the material of one picture or one line of text, to something like organized information. Take news as an example. When we use the word *software* we often refer to the entire news program itself, whereas when we use the word *contents* we can also be referring to the images and illustrations that compose the news program.
>
> However, why is that we chose not to use the word *software*—one that we have become familiar with—and instead use the term *contents*? The reason for this lies in the existence of multimedia. Within the environment of multimedia, we engage not only with software itself, but also with information exchange which occurs at the level of the materials that compose the software, such as voice, image and so on. As a result, the unit of information shifts from software, to contents.[53]

Here, then, the contributors to *Jōhōka hakusho 1997* attempt to work through what was for them the privileged term until recently—*multimedia*—the subsequent term, *software,* and the new kid on the block, *contents.* Their answer suggests contents is a third term, not simply a replacement for software, or multimedia, designating a smaller or modular unit of information.

Indeed, the authors suggest something like a generalized state of multimedia, in which there are larger blocks called software (or, alternatively sometimes, services), under which there are smaller composing elements called contents. Their suggestion is that—to put it in a somewhat confusing manner—contents could reference the content (component parts) of software. This points to the ambiguity of the term *contents* that could refer to a particular larger-scale entity like a video game (much as software or *sofuto* did), or it could refer to the content of this *kontentsu,* which is to say one particular scene in a game, or a character, or a sequence of the soundtrack. Perhaps fittingly for a white paper concerned with informatization, the problem of contents came down to the scale of the information, or what we might call the size of the information block. Contents had a certain ambiguity, wherein it could refer to something large-scale like software, or it could refer to the smaller-scale building blocks (image and sound "content") of this software.

Finally, in a position that reflects the "content is king" mentality of the 1990s, the *Jōhōka hakusho* contributors suggest that unlike the predominance of media and multimedia in former days, at present (that is, in 1997), contents was increasingly more important than media. In an era when there were distinct media, it was the media that had rarity value, and contents accumulated around them. However, in a multimedia age, the rarity value of media is in decline, and what becomes most valued is contents. This, the authors declare, is the age in which they find themselves. An era when contents is clearly king—even if they do not trot the phrase out.[54] This means that the topic of copyright is an essential element of the contents business; for "in the age of multimedia, one contents can be developed across numerous media"—a classic formulation of the connection between contents, multimedia, and what is by then called the media mix.[55]

Here I would like to turn to a third white paper, *Jōhō media hakusho* (Informational media white paper) (which comes with the English subtitle "A Research for Information and Media Society"), edited by Dentsū Sōken, a research division of Japan's largest advertising agency, Dentsū (about which I will have some more to say below). Perhaps being associated with the ad firm, and hence much more trend-sensitive, the editors of this white paper acknowledge the term *contents* in their 1996 (published in January 1996) edition of the white paper, a year and a half earlier than the more conservative and hardware-oriented *Jōhōka hakusho*. Here too, the strange relation between soft *(sofuto)* and contents *(kontentsu)* is the crux of the problem. In an opening section called "How to Use This Book," the editors helpfully define their terms. There they introduce the term *contents* in the second paragraph, in the following manner: "The information industries can be divided into the information machines industry, the information services industry, and others that allow one to use soft (contents). In the narrow definition of the term, information media industries have the information services sector at the center."[56] They continue specifying the meaning of *contents* in a section—titled "Media and Soft"—that also defines the ambiguous term *soft*.

> According to the general industry divisions, the software industry would be considered part of the service industries. Soft is the abbreviated version of software—as opposed to hardware—and emerges as a specialized term from the field of computers. Since in the case of computers soft often refers to "application soft" (basic control soft), it is common to refer to what would be considered information soft as "contents." However, generally what is called soft refers to the informational content *[jōhō naiyō]* recorded (or stored) on media (the medium or physical media) of video tapes and CDs.[57]

This definition usefully suggests the semantic slide whereby a term that had formerly referred to the recording medium (VHS tape) by 1996 referred to both the physical substrate (video tapes or video game cartridges) and the information content of the tapes or cartridges—which is to say, the media product in question, whether *Top Gun* or *The Legend of Zelda*.

By the 1998 edition of the *Jōhō media hakusho, contents* is newly glossed as a more recent word for *soft*.

> The "soft" (recently it has become common to term this *contents*) within information media is composed by taking components such as "data" *[jōhō]* and "information" *[infomēshon]* and collecting/editing/reprocessing it into a higher order form which is "soft." This soft is subsequently copied (by printing, record presses, and so on) and then distributed to large numbers of people who then use it.[58]

While some of this information is combined in digital form as "multimedia," the "largest proportion of the information industries are separated into the particular expressive forms of film, publishing, records, games, and so on."[59] Film, publishing, and other industries have within this account become branches of the information industries, and soft is something equivalent to media, composed of lower-level information. The reconfiguration of the film, music, and book publishing sectors into subsets of the information industry, under the aegis of the keyword *kontentsu,* appears complete. It is in this context that we can better understand why Kadokawa Tsuguhiko could, several years later, insist that his publishing, film, and animation company could be a "mega contents provider." For a full picture of the importance of the plural form of *contents* within the Japanese media industries, I will now examine the relation between contents and the media mix.

From Narrative Marketing to Contents Marketing

The first section of this chapter followed the semantic transformations of contents as a keyword within digital culture, tracking how it both replaces medium-specific terms like film or video and displaces earlier terms like the already ambiguous and polysemantic "sofuto." Here I turn to another part of the story of *kontentsu* in Japan, its metamorphosis in the twenty-first century, coming to designate anime, manga, light novels, games, and other forms of Japanese "subcultural" media.[60] In part under the direction of METI, and in part through the existing emphasis on animation and game soft, contents itself becomes a term that designates these particular genres of media above others. The uniqueness of the term *contents* in the Japanese context is that it becomes associated with particular kinds of media (unlike the English term, whose defining characteristic and indeed its value lie in it being a medium-agnostic term). The ambivalent, dual meaning of *kontentsu* is illustrated by Hatakeyama Kenji in his prologue to the influential 2005 book *Odoru kontentsu bijinesu no mirai* (Future of chaotic entertainment content biz), which he coauthored with the executive producer of Pokémon anime films and TV series, Kubo Masakazu. Here, Hatakeyama sets out to offer a definition of the "contents business"—a term he shows to be rather elusive.

> Since we use the word business, contents must certainly be the business of making deals. If so, then what is contents? In newspapers and magazines, the word contents is frequently used interchangeably with the terms "informational substance *[nakami]*" or "informational content *[naiyō]*." It is true that the word *content* [English singular in original] is used to mean content *[naiyō]* or substance *[nakami]*. . . .

On the other hand, we are told that contents is anime and film, music, video games and so on.

So informational substance equals anime, film, and so on? What is this?![61]

Hatakeyama never answers his own question. Instead, he returns to where he started, concluding that "we call film, anime, comics, video games, music and so on contents, and call the business of making deals with these things the contents business."[62] Echoing the rhetoric of the Cool Japan strategy (already in high gear by this time), Hatakeyama concludes on a saccharine note by adding that "'contents business' is a word that Japan can proudly transmit to the world."[63]

Hatakeyama's ultimate evasion of the question he poses leaves something to be desired. Help might come from turning our attention to the realm from which the connections between the specific media of anime-manga-games with the general term *contents:* the realm of marketing discourse, and discussions of *narrative marketing* in particular. The wager here is that the most satisfying answer to the question, "What is contents?"—the question we must answer before turning to that other question that animates this book, "What is a platform?"—can be arrived at by examining the predigital practice of marketing multiple forms of related cultural commodities through the media franchising strategy called the media mix, and its theoretical formulation under the moniker of narrative marketing. The subsequent shift from terming this practice narrative marketing to calling it *contents marketing* makes visible the dependence of the latter on the media mix model, and the incorporation of the media mix into contents industry practice. If hardware and software industries are one place where we find the term *contents* emerge and take form, another is to be found within the media industries, and in narrative marketing techniques built around them. Following the shift from narrative marketing to contents marketing allows us to track the discursive contours and definitional forms of this keyword; it is a site from which we can observe how the now almost transparent term *contents* rides on existing media and discursive formations, even as it transforms them. As we will see shortly, the proponents of narrative marketing successfully rebranded this practice contents marketing in the early twenty-first century. The substitution at play in this rebranding provides the basis for some insight into the privileged relationship between narrative and contents.

Crucial here is the Japanese transliteration's emphasis on contents in the plural, and its implication that there is always a multiplicity of contents. In this sense it is no coincidence that the media at the locus of transmedia convergence—anime, manga, light novels, games—are the very same media most often associated with the term *contents*. Contents here is not quite as "medium agnostic" as its North American interpreters claim; rather, in Japan's case the term designates the media genres most associated with the practice of media convergence or transmedia storytelling known as the media mix.[64]

Contents is arguably the buzzword that best captures the phenomenon of the media mix, increasingly replacing the latter term over the course of the early twenty-first century, subsuming the media mix into contents business practice. (That both terms start as marketing or business world lingo should also not escape notice.) The plural form of the term is not an error of transliteration but rather demonstrates how *kontentsu*/contents as discourse assumes symbiotic networks of media and cultural commodities that traverse from one platform to another, a mobility aided and abetted by the increasingly medium-agnostic condition. Hence the media mix—as the cross-platform movement of texts—is a fundamental building block of the contents business that Hatakeyama and others focus on. In a circular manner appropriate to the its guiding principle of "synergy," the media mix as a marketing practice is both a condition of possibility for the field of contents business and the target area that this business hopes to mine. It is no coincidence that the cocreator of Pokémon, a massive global media mix success, is the coauthor of one of the most impactful contents business books in the first decade of the century.

To best demonstrate this will require a short detour through the historical development of narrative marketing as a subset of marketing theory and practice. To do this I will briefly suspend my discussion of contents, focusing instead on the Japanese history of advertising. This will, ultimately, lead back to that account of contents and contents marketing in particular—but in order to see how contents plays out in advertising discourse, I will take a short detour through precontents Japanese advertising history.

From Needs to Signs

Histories of Japanese advertising emphasize a historical rupture around the 1960s, at which point the logic of marketing shifted from an emphasis on the use value of commodities to an emphasis on their symbolic or sign value.[65] This narrative often demarcates two distinct periods: first, an era of marketing based around needs; and second, an era of marketing based around affect, the play of signs, or ads without necessary reference to the things advertised. To this, building on the work of such scholars as Ōtsuka Eiji and Fukuda Toshihiko, we can add a third era: a time after the exhaustion with the play of signs, when marketers turn to narrative (narrative-based ads and narrative commodities and spaces) in order to incite consumer desire.[66] While it is unlikely that there was ever a moment where marketing was based purely on needs, and the narrative is often told by the makers of the ads themselves, this narrative is convincing when considered in light of a review of commercials during this period, which do evidence a thoroughgoing change in advertisement and commercial style from the 1970s onward.[67]

The first period begins sometime during the immediate postwar, a vaguely defined period roughly from the mid-1940s to the late 1960s during which commercials spoke eloquently of a particular commodity's usefulness, its attributes, and its functions. Ads appealed to the consumer's good sense in making their sales pitch. The basis of a commodity's appeal lay in its use value, and the commercial was a means of demonstrating a commodity's usefulness. This involved a sometimes elaborate display of its functionality, and its quality of construction. A car commercial would emphasize a car's speed, or longevity; a commercial for a vacuum cleaner would demonstrate how well it cleans, or the convenience afforded by the location of the on/off switch.

Then, at a certain point in time, usually located in the very late 1960s or early 1970s, the market became saturated with commodities that were indistinguishable in terms of their level of quality. One vacuum cleaner was really no different from another. The quality or durability of an item could no longer be its selling point. Moreover, consumers had reached a point where their basic living needs were fulfilled. Most had attained the middle-class lifestyle so actively promoted; they had also acquired most of the commodity durables that had become the measures of middle-class belonging. For the first time since the austerities of the Pacific War and the material poverty that followed, a level of consumer satisfaction and commodity saturation had been attained. Marketers could no longer operate on the premise that the consumer knew which commodities were necessary for a lifestyle on par with that of their peers, and merely needed to be convinced that the product or make in question was the one to go with. The conclusion was that needs had to be created and could no longer be assumed. With this conclusion, Japanese marketing entered the second period noted above, what we might call a "postutilitarian age."

In the postutilitarian age, needs had to be created in new ways. Admen and women could no longer appeal merely to a consumer's good sense in making a choice between better and lesser commodities. Emphasis on the functions of a particular commodity gave way to emphasis on feeling, lifestyle, and other qualities that did not so much distinguish one commodity from another as mark one commodity with a particular affect, quality, or sensation. The shift, usually dated to the years 1969 and 1970, implies a shift in the production of commodities themselves. As Kojima Tsuneharu notes in his overview of the relationship between commercials and marketing, whereas prior to this time "the development of new products were centered around the development of new *functions*," from 1970 onward "the development of *value* increasingly became a central theme."[68] This shift in emphasis from functions to value was understood as a shift from "intense shoppers" born in the prewar or wartime periods to the increasing dominance of a class of consumers born in the postwar period, dubbed "beautiful shoppers" by Kojima, following a 1974 survey on women's behavior and mindset.[69]

This was a shift from an age of use or need to an age of sensibility or feeling. One prime advocate and narrator of this shift was Fujioka Wakao. Fujioka was a creator whose work would come to define an era; he worked for Japan's largest ad firm, Dentsū (the research department of which would, as we have seen above, go on to be an early adopter of contents discourse in its 1990s white papers). Fujioka's famous "Discover Japan" campaign of the 1970s is a landmark in Japanese advertising history, a defining moment in postwar Japan, and the subject of an excellent analysis by Marilyn Ivy.[70] But in what is perhaps his equally famous 1970 campaign for Xerox— analyzed by Tomiko Yoda—he coined a copy phrase that would stand in for a massive historical shift: *"Moretsu kara biutifuru e"* (From intense to beautiful).[71] *Beautiful*—in its English transliteration rather than the Japanese *utsukushii*—evoked the hippy era of love and peace, signaling a different approach to work, the environment, and life. The campaign marked an epochal shift. In its representative product, a thirty-second television commercial, the camera keeps pace with folk singer Katō Kazuhiko as he walks down a busy Tōkyō street, dressed in hip clothes and carrying a handwritten sign that reads "BEAUTIFUL."

The image is blurred, particularly around the edges, and is marked by lens flare. On the soundtrack is a song sung by Katō, whose lyrics are simply the word *beautiful* repeated over and over. Suddenly the sound cuts out and the image freezes on Katō, as the words *Moretsu kara biutifuru e* appear over the image, cutting then to a white frame with the word *XEROX* in the center. The ad differs remarkably from earlier Xerox ads, which emphasize the speed and efficiency of Xerox photocopiers, whose copying speed is "as fast as the wind," as a 1968 commercial puts it; or, as in another commercial from 1960, offer a quick visual demonstration of how particular copiers (in this case the Xerox model 660) work, concluding with the tag line "Xerox, the new rhythm of business."

In stark distinction from these earlier commercials, which display the ease of operation and functionality of Xerox, the 1970 "Beautiful" ad presents Xerox as offering not a new commodity but a new lifestyle, a change of the very way of life from the hard-working Intense to the quality of life of Beautiful. As such this ad more than anything represented a shift in advertising itself—from an era of ads that worked hard to convince people to buy their products to an era where ads worked through feeling. Indeed, it was Fujioka himself who most clearly articulated this shift in his 1984 book, *Sayōnara, taishū* (Goodbye, masses), in which he narrated the end of mass culture and the rise of the *shōshū*, or "micro-masses."[72]

This was also the era of what Fujioka and others called *"monobanare kōkoku"* (ads separated from things), ads that are no longer concerned with the things they are selling.[73] As the specificity of the products began to disappear, and there were "no longer any particularities or sales points to the products themselves," ads "were forced to separate from the product message, gave up praising the product, and progressed by distinguishing themselves at the level of [ad] expression alone."[74] The representative example of this new form of advertising was the "From intense to beautiful" Xerox campaign, a form of *"datsu-kōkoku"* (deadvertising) in which ads no longer enumerated the qualities of a product in question but merely functioned as autonomous forms of expression that referred only to themselves.[75] The Xerox ad is the perfect example of this, insofar as it is impossible to tell exactly what the ad is promoting. We cannot see a single example of a photocopied object (much less a photocopier) throughout the entire commercial. Feeling is the order of the day, not information—albeit to a perceptive viewer the repetition of the single word *beautiful* may indeed function to conjure up the copying effects of Xerox's machines. Much like the Xerox example, ads hereafter functioned to mark arbitrary differences between commodities or, in more attenuated form, merely marked differences between ads themselves.[76]

The 1980s sees the height of consumption as guided by the play of signs, a marketing practice inspired in no small part by the influx of structuralist and poststructuralist critical theory. Jean Baudrillard's early works *The Consumer Society* and *The System of Objects* were both translated and eaten up by ad executives and academics alike. Semiotic theories of consumption became the rage, and the subsequent generation of academics of the 1980s was supported in large part by the interest of ad agencies in functionalizing theory. Dentsū, for instance, supported writers such as poststructural theorist Asada Akira; feminist critic and scholar Ueno Chizuko; manga scriptwriter, editor, and writer Ōtsuka Eiji; and feminist sociologist Kayama Rika. Raised on poststructuralism, Asada and other New Academic theorists translated Deleuze and Guattari's schizoanalysis into a kind of social theory as advertising theory.[77] There was a collapse, that is, between the method of cultural analysis and cultural production. The "play of differences" became at once a poststructuralist motto and the principle of advertising of the 1980s. Critique, then, was mobilized for advertising praxis.

Yet difference in and of itself could only work as marketing strategy for so long before generalized disorientation, fatigue with the play of signs, and sign-induced apathy began to take over. Affect and signs alone were not enough to induce consumers to pick particular products. At this moment a few marketers stepped forward to advocate a return to some level of grounding—this time in narrative. Marketing theorists began advocating a (re)turn to narrative as a means of hooking in consumers exhausted by the infinite play of differences—bringing us to the third stage in the narrative of Japanese postwar advertising method, the era of narrative marketing, which in turn takes us one step closer to contents discourse.

From Signs to Narrative Marketing

By the late 1980s and early 1990s, marketing commentators such as Inamasu Tatsuo noted that the proliferation of signs and the "dismantling of narrative" of the 1980s had given way to a "rebirth of narrative" in the 1990s.[78] At this moment, there were at least two groups of persons dedicated to investigating the potential for mobilizing narrative for the purpose of attracting and maintaining the interest of consumers in particular products. The new field was known as *narrative marketing,* and the two main players investigating it in the late 1980s and early 1990s were Ōtsuka Eiji, who worked as editor, manga scriptwriter, and media mix producer for the Kadokawa Media Office division of Kadokawa Shoten (Kadokawa Books; later Kadokawa); and Fukuda Toshihiko, a marketing researcher who worked for Dentsū. Each acknowledges the other in their work; Ōtsuka refers to the interest in narrative marketing or storytelling marketing among a group of ad researchers at Dentsū, a group whose core member was Fukuda.[79] And Fukuda acknowledges both the stimulus received from Ōtsuka's work, which was being serially published in, among other places, the Dentsū in-house journal, *Dentsūhō,* and the overlap between Ōtsuka's ideas and his own.

Today, Ōtsuka is renowned as a theorist of manga and anime, a scriptwriter for manga, a novelist, and a writer of how-to guides for writing novels that moonlight as literary theory and criticism. But in the late 1980s and early 1990s he was also working closely with a media division of the publishing giant Kadokawa Shoten. There, Ōtsuka worked under Kadokawa Tsuguhiko within the Kadokawa Media Office (later Media Works), which was developing a new model of the media mix geared toward small market segments and inspired by the open narratives of the then-popular Table Top Role-Playing Games (TRPGs). It was in this context and at this time that Ōtsuka published his first books. The most prominent of his early works included an ethnographic account of narrative consumption patterns of children of the late 1980s, *Monogatari shōhiron* (A theory of narrative consumption).[80] While ostensibly an urban ethnography of the narrative consumption of children, this book was also, as Ōtsuka himself emphasizes, a work aimed at developing a practical model of narrative consumption, a new model of the media mix, and a strategy book for Kadokawa Media Office. At one point, Ōtsuka calls the book "marketing theory written in order to make inroads into Dentsū and Kadokawa Shoten."[81] Elsewhere, Ōtsuka notes that the book was at the time mostly read by marketing executives. *A Theory of Narrative Consumption* was, Ōtsuka writes, nothing less than "marketing theory from the 1980s."[82] While I have previously examined this development in *Anime's Media Mix,* I would like to briefly return to it here to situate it within the history of advertising, noting its role in the rise of narrative marketing.

The core of Ōtsuka's analysis in *A Theory of Narrative Consumption* was the phenomenal success of "Bikkuriman Chocolates"—a kind of chocolate candy that came with sticker premiums or freebees bearing the image of, and information for, a number of different characters. What set Bikkuriman Chocolates apart from earlier premium-based campaigns was that each sticker bore a fragment of a larger narrative that children could only piece together through the consumption of these very stickers. As Ōtsuka suggests in his analysis, each sticker functioned like one episode in an anime series; only by watching the entire series—or in this case collecting it—could one grasp the totality or grand narrative that informs it. The consumption of each fragment led to a greater desire to piece together the narrative whole. The genius of Bikkuriman Chocolates was to have used the sticker premium as a medium for the development of the narrative. Buying the chocolates was the only way children could mentally reconstruct the mythic narrative of which each sticker supplied a fragment. This reconstruction made it possible for the consumer to eventually participate in the creation of the narrative. With this return to narrative, Ōtsuka detected a way of getting consumers hooked, a way that lead beyond the mere play of signs of high poststructuralist thought and the resulting consumer fatigue. For what better tool was there than narrative to give meaning to the supposed arbitrary array of signs endemic to the consumer landscape of the late 1980s?

Narrative marketing was a highly effective means of attracting consumers in an age of sign-induced exhaustion. If Ōtsuka was one advocate of narrative consumption, his contemporary, Fukuda Toshihiko, became the writer and researcher and marketer most associated with the development of narrative marketing—working from within the belly of the Dentsū beast. Fukuda's 1990 book *Monogatari māketingu* (Narrative marketing) opens with a similar premise to that found in Ōtsuka's essay, and to post-1970s marketing in general: the shift from need or use value to sign value as the principal motive for consumption, and the logic of advertising that guides it. What Fukuda adds is that in a commercial environment characterized by the overproliferation of signs, what gives these sign commodities meaning is narrative. Narrative is one way of producing desires in an era when an appeal to use value is no longer possible, and sign value has lost its allure. Narrative produces value and fabricates needs. As Fukuda writes on the opening page of *Narrative Marketing*:

> In contemporary society, things and information are overflowing far beyond the level of biological or economic "need." In this "society of overabundance" the consumer can no longer feel enticed simply by high functionality or high quality goods and services. This is a state of affairs wherein things like comfort and playfulness no longer have an impact.
>
> Within this age, product development, store development and promotions that use narrative as a key are gaining attention as being effective marketing techniques.[83]

The future of marketing lies not in things *(mono)* but rather, in Fukuda's pun, narrating things *(mono-gatari)*. People have tired of consuming things as signs; they need new motivations to consume. Narrative marketing promises to create this motivation, and possibly retain consumers for longer periods of time.[84]

Narrative is effective as a marketing tool because it is a fundamental part of human experience. Stories, says Fukuda, create the frame through which humans experience the world; "narratives are eyeglasses we put over our psyche in order to understand the world."[85] Stories are a kind of Kantian schema that is narrative in form; narrative filters and makes legible human experience. Given his fervent belief in the importance of narrative in structuring human experience, it is no surprise that Fukuda latches onto narrative marketing as the perfect tool for need creation, and for sustaining consumption beyond the play of differences.[86] The play of differences in commodity form was all well and good, but true to its Derridean coinage, the play of signs was unstable, and the commodities characterized by this model quickly lost their appeal. Storytelling, on the other hand, would be a way to create continuity between commodities, and to prolong the longevity of a given commodity.[87]

How Narrative Marketing Becomes Contents Marketing

When Fukuda first published his book in the 1990s, he found some resistance to his suggestions. In the introduction to his 1990 book, Fukuda writes that narrative marketing is a new approach that does not yet exist in practice.[88] It appears that narrative marketing had some difficulty catching on during the 1990s, at least within the mainstream of marketing, where many regarded Fukuda's work as heretical.[89] Quite the opposite was the case for Ōtsuka's work with Kadokawa Shoten, which gained prominence over the course of the 1990s and into the new millennium, such that Kadokawa, as the new conglomerated book publisher and internet company, has become an increasingly significant player in both print and video, and increasingly in games as well. Despite this initial disdain in the realm of marketing, narrative strategies have seen a small renaissance in recent years, particularly in light of the growing public and governmental importance placed on the narrative media of anime, manga, and video games. This resurgence of interest coincides with the emergence of the term we traced in the first section of this chapter, and that has replaced narrative in importance: *contents*.

Nothing reflects the rise of the term *contents* better than a 2004 book Fukuda coauthored with younger marketing scholars Arai Noriko and Yamakawa Satoru, titled *Contents Marketing: Searching for the Laws of the Narrative Commodity Market*.[90] The book is fascinating for its close relation to Fukuda's earlier volume, *Narrative Marketing*. Indeed, in some ways this book is a more breezily written recap of Fukuda's 1990 book, aimed at a popular readership: a kind of how-to manual for narrative marketing. Yet it is in the very similarity between these two books that we can find the basis for making sense of the new term introduced in its title: *contents.* Whereas the title of Fukuda's earlier work was *Narrative Marketing,* the more recent coauthored volume offers up a parallel title, *Contents Marketing.* In this book and so many of those volumes from the early twenty-first century onward that consider anime, manga, or games from a business perspective, the term *contents* seems to be merely a newer term for narrative, or more precisely, *narrative-based commercial media.*[91] Just as anime, novels, games, and films are the privileged types of contents (albeit with the requisite mention of music or social media), for Fukuda, Arai, and Yamakawa, contents is merely a stand-in for the older term *narrative*—hence the easy shift from narrative marketing to contents marketing. But just as narrative for Fukuda is a mere schema for meaning making, contents too can be said to function as a schema, this time for value creation in the digital world, at a moment when the dematerialization of media threatened the package-based business model of the time.

As Thomas Lamarre writes, in a different context but in a passage that reframes the function of contents in an important way: "We see concentration of ownership, proliferation of channels and independent producers, and transnational serialization—a licensed repurposing of contents, in which contents have been transformed into a sort of schema or concept rather than contents in the usual sense."[92] Contents, then, is a "schema" rather than a mere substance. It is a form of packaging, a filter that endows entertainment goods with economic value. And yet picking up this issue of the transnational serialization of contents, we should reopen the issue of the plural form this term takes, the curious fact of the plural form *kontentsu* in the Japanese context. Why is contents plural and not singular *content?* Here too Arai, Fukuda, and Yamakawa offer some guidance.

Unlike the government white papers that retranslate *kontentsu* as content in the singular, Arai, Fukuda, and Yamakawa note and understand the significance of the plural form of *contents.*[93] The object of contents marketing is not a single object, service, or experience (a content) but is rather the multiple contents, that is to say the media mix formation itself. In a chapter titled "Contents Research Is the Cutting Edge of Marketing," they write:

What we aim to do is not simply to market a single work (commodity); rather we aim to strategize how to develop markets based around the ripple effects of this work. Thinking "How should we sell this film?" involves not only trying to work at increasing the audience for the film, but also thinking about how to sell the original work as a novel or soundtrack, how the theme song might be downloaded by the largest number of people for use as their ringtones, how some of the film's dialogue might turn into buzz words—in other words, thinking about how to create a social phenomenon out of a single contents. We can say that the whole concept of today's contents business is based around a total return that incorporates the ripple effect.[94]

The ripple effect, the interconnection of cultural commodities, and the goal of a synergetic whole all point to the phenomenon of the media mix. Indeed, Arai, Fukuda, and Yamakawa make explicit reference to Kadokawa Haruki's media strategy, often considered as the origins of the media mix in Japan, which they describe using the 1980s term "Kadokawa shōhō," or Kadokawa commercial strategy. They then redescribe this Kadokawa commercial strategy as a "methodology for the 'expansion of contents' from an original work to other formats."[95] What was formerly called the media mix (or the Kadokawa commercial strategy) is newly dubbed the "contents mix" *(kontentsu mikkusu).*[96] This formulation recalls another popular term that mobilizes the term under consideration here, and operates as a stand-in for the media mix: "one contents multi use" *(wan kontentsu maruchi yūsu),* a term we find deployed in many of the books considering contents.[97]

Contents is pluralized because it is always multiple and iterated across media forms. *Contents is always already media mixed.* Contents is plural because the cultural goods packaged by it are serially connected, within a media environment where connections between disparate media are the norm. Contents as a term gestures at this fundamental plurality: content is never singular but rather always comes in series, and always is one step away from the next media transformation. In the Japanese context, content is always contents. Arai, Fukuda, and Yamakawa's *Contents Marketing* reminds us that the real referent of contents, and the reason for the term's plural form, is the multiplicity of media and their system of relations known as the media mix.

On the Function of Contents

Moving from the above account of marketing's use of the term *contents* to a critical diagnosis of the function of contents as term, as discourse, and as a worldmaking practice, I return in this concluding section to Lamarre's point that contents are a schema. In what follows I move from discursive analysis to wider speculation about the function of contents, as discourse and as manner of managing media practice. Indeed, ultimately my claim is that contents functions as a form of discursive and economic packaging that endows cultural entertainment goods with economic value, preparing them for the platform intermediation that is the subject of subsequent chapters.

The first part of the contents story told above charted the conditions for the emergence of the term: on the one hand, the dematerialization of media with the increasing dominance of medium-agnostic digital media, and on the other hand, a parallel lineage wherein contents replaces the similarly ambiguous term *sofuto*. The second part of the contents story was an account of the mysterious semantic layerings of the term, wherein contents designates privileged media forms: anime-manga-games-novels and, by extension, the associated media mix practice in toto. Indeed, just as anime never functions alone but is always part of the media mix ecology, so contents never functions in the singular but always in its synergetically linked plural form.

To complete the world-impacting and industry-changing implications of the rise of the keyword *contents,* I would like to wrap up this chapter by considering function of contents discourse as it impacts contemporary Japanese media industries. In what follows I pivot from the more careful tracking of the term as discursive formation to speculation about the reasons for the term's sudden proliferation in Japan (not to mention other Asian contexts such as South Korea), its role as panacea, and its impact on contemporary media industries.

Much as narrative operated as a panacea to companies in search of a new kind of marketing in an era of exhaustion with the play of signs, contents arrives on the scene at the precise time of anxieties around the rise of digital media and the potential dematerialization of audiovisual commodities. If, for Fukuda, narrative gives form and meaning to our world, the term *contents* works in a parallel manner, giving value form and economic meaning to the digital world. To go a step further: contents-speak is a tool for discursively packaging cultural goods within commodity forms in the era of digitally mediated transmedia consumption. If the model for media production is no longer a discrete commodity (a book) but rather a transmedia commodity array (book-anime-game-toy, and so forth), a word is needed to describe this medium-agnostic sequence. Contents seems to be the ideal candidate in its ambiguity, openness, and relative inoffensiveness (much better than, say, the lawyerly sounding "intellectual property," or IP).

Contents is a first step in creating value out of digital data, making data into goods. Platforms follow up this magic trick with hardware- and software-based digital rights tools that limit contents' circulation (thereby creating the "walled garden"), and transactional mechanisms that support the creation and valuation of contents. Hence the trick contents performs is to give bounded form to cultural commodities in an era of digital media. Alongside platform-based digital locks and digital rights management tools, this imagined or discursively produced boundedness also offers a sense of reassurance to IP holders who fear the sometimes overly fluid and facile copying and circulation brought on by digital media's lower barrier of entry, and the minimal cost of copying. As economists Carl Shapiro and Hal Varian noted in their 1999 book, *Information Rules,* "Information is costly to produce but cheap to reproduce"—and it was this low cost of reproduction that most scared IP holders.[98] *Contents* as a keyword creates the promise that a given cultural good could be monetized. *Contents* as narrative may give form to the experience of our world (pace Fukuda), but *contents* as a term is a schema that gives economic form and value to cultural goods.

This is why contents discourse is so strongly tied to business literature, governmental policy, and how-to books on value creation, even as it has subsequently become the topic of academic inquiry, and everyday parlance. Contents discourse hence operates as a discursive panacea to the perceived danger of digital media and the threats to structures of monetization of cultural goods that were still very much tied to the purchase of material goods such as CDs, Blu-ray discs, and books. The proliferation of contents business books promises the holy grail of revenue flow in an environment of supposed digital dematerialization. To repeat, contents discourse plays the same role as narrative does in Fukuda's account of narrative marketing: *it gives form to an unsettlingly amorphous world.* (It also offers the promise of a potentially lucrative new market for hardware companies seeing a drop in revenue with the commoditization of the hardware sector in the 1980s and 1990s—the reason for Fujitsu's venture into contents, and its key role in the popularization of the term.) Through contents, the world of cultural goods becomes reassuringly legible to industry players as commodities. The form contents gives the world is, perhaps unsurprisingly, an old one: the commodity form.[99]

Contents is a kind of discursive parallel to packaging, somewhat akin to the plastic packaging that wraps goods and imparts them with the commodity form. Susan Willis rightly suggested that while Marx started his analysis with the commodity, we should instead start our analysis with the "commodity as it is metaphorically reiterated in its packaging."[100] Contents is a discursive re-formation of the postconvergence cultural product as commodity, as something contained in a form that imbues it with economic value. As Sumanth Gopinath writes in the context of ringtones, "In the lingo of the [ringtone] industry, 'content providers' are those companies and institutions that offer the entertainment substance of what is being sold."[101] Contents is the commodity formalization, the packaging of this entertainment substance. And it bears reminding that ringtones, as we will see later in this book, are one of the first contents of Japan's megaplatform, i-mode.

To conclude, allow me to sum up the findings of this chapter in relation to the stages of adoption of the term *contents.* First, the term *contents* came into use in Japan in the mid-1990s in order to designate something like informational content, data, or media in an era of increasing digitization and the attendant transferability or transposition of files or media formats from one platform to another. It also displaces the existing keyword of the 1980s and 1990s for video tapes and game cartridges and other media: *sofuto* (software). Second, by the early twenty-first century, the term *contents* starts to reference particular genres of media, namely anime, manga, games, film, and other cultural commodities that were products of Japan's creative industries and seen as capable of export under the then-new "Cool Japan" label. At this point, the term takes on the sense of packaged intellectual property, offering both the sense of a discrete entity—a Blu-ray disc—and the transmedia extensions of this entity within the media mix. Third, and finally, running parallel to this development of contents discourse is the theorization of narrative marketing or narrative consumption that emerges in the late 1980s and early 1990s from developments in marketing. The importance of narrative marketing comes on the one hand from its close relationship to the media mix, with the work of Ōtsuka and others at Kadokawa Media Office, and on the other with Fukuda's recent rebranding and reconceptualization of narrative marketing as contents marketing. Arai, Fukuda, and Yamakawa's work allows us to see that contents functions as a form-giving and value-endowing discourse and practice. As we will see in subsequent chapters, it emerges slightly prior to platform theory and platform building in Japan. That said, the platform quickly becomes the basis upon which contents take form, and are contained in economic value.

Contents discourse in Japan is hence configured as this unique combination of (1) contents as information or medium-agnostic media forms; (2) contents as anime-manga-games-film, as packaged, exportable IP; and (3) contents as a particular media mix formation, in the age when the very term *media* seems on the verge of disappearance. Contents in Japan is hence never configured as a singular content—a manner of thinking that Bharat Anand in a recent business book calls "the content trap."[102] Rather, contents in their multiplicity are already involved in the synergies and network effects that mark platforms. Moreover, the very possibility of selling contents depends on the bounding properties of platforms, the operational closure of "walled gardens" that prevent the free flow of digital files and allow for the monetization of digital content—ringtones and beyond. A strong platform—as we will see later in this book—will in turn create the technological conditions and market conditions for the packaging and selling of cultural goods as contents, a value-creating moment that the media industries live for.

CHAPTER TWO

Platform Typology

From Hardware to Contents

*C*ontents as a keyword in Japan and the United States saw a meteoric rise during the 1990s and into the early twenty-first century, cementing its status with the now famous moniker Content Is King. Contents (in the always plural Japanese form *kontentsu*) *was* king in Japan, from the mid-1990s through the first two decades of the twenty-first century. Film companies became contents producers and book publishers mega contents providers; and the government's Cool Japan campaign was all about the country's coolest contents: anime, games, manga, light novels, refigured under the global category of contents. Contents as term functioned as the discursive packaging of dematerialized digital goods ("data") as commodity forms. Platforms, in turn, are the infrastructural and transactional basis for value creation.

The priority given to contents or platforms respectively seesaws back and forth depending on the era. During the 1990s in the United States, contents was king, as famously proclaimed by Bill Gates. However, by about 2005 changes were afoot. Content was no longer king. Instead, the attention shifted to something else, something that would come to be called *platform*. Indeed, it would only be once platforms were in place by the end of the century's first decade that the future of the internet as broadcasting channel would be secured. The future dreamed by Bill Gates and AOL Time Warner, which aimed to merge content production with platform delivery, only arrived by the 2010s as Apple, Google, Amazon, Netflix, and other platforms ecosystems unequivocally announced: platform is king. By this time, Apple, Google, and Amazon were engaged in a fight for national and global platform dominance, particularly via their smartphone platform proxies.[1] Platforms as sites of content containment, distribution, and management now seem king, with the *intermediaries of content*—rather than producers or owners—the ones increasingly celebrated (or, alternatively, vilified) within the media landscape. Google and Apple and Amazon have developed closed platforms as contents-delivery ecosystems that sell and distribute existing contents. Emphasis falls on the creation of platforms that open onto closed ecosystems of contents, with the internet functioning as an architecture of diffusion and transmission.[2] With the rise of Netflix and its contents-driven strategy over the 2010s, the balance seems to be shifting somewhat toward *contents,* with platform giants now entering the content production game. Yet Neflix itself is a platform first, and a content producer second.

These periodic seesaws in terminological and business dominance are equally visible in Japan. In the United States, as I noted, the platform was becoming king in the early twenty-first century. In Japan during the same period, contents appears to be king. At first glance, more seems to have been said and done about contents than platforms. And yet, once we scratch the surface of contents discourse in Japan, we find that the basis for the rise of contents in the Japanese media sphere was the incredibly robust platform systems in place. As we will see in the coming chapters, there has been consistent and considerable engagement with platforms in Japan from the mid-1990s onward, culminating in one of the key technological systems and first incarnations of the platform in its current form as both technological set-up and economic intermediary—in the form of i-mode and the other mobile internet systems. Scratch below the surface and we find that Japan was one of the platform capitals of the world in the late 1990s and the following decade, and this platform power in turn provided the basis for the contents bubble that operated at both financial and discursive registers from about 2000 to around 2005–6.[3] It also—arguably— drove U.S. industry giants such as Google and Apple and Amazon to take up the cause of platforms, modeling their mobile platforms on those popularized in the late 1990s in Japan.

Platforms: Toward a Historical Typology

But what is a platform? Despite the considerable amount of attention paid to platforms, and a good deal of programming and planning put into physically building platform ecosystems, there is not a consensus on what is meant by *platform*. Indeed, even as the use of the term is on the rise, there are competing definitions of the term, often presented as incompatible with the others.

This chapter offers an overview of the most prominent of these competing definitions, in relation to both the English-language (predominantly North American) and Japanese-language contexts. Excellent overviews of the term already exist within management literature and more recently within critical theory as well.[4] Yet these tend to exclude media studies in their focus on management, or conversely exclude management studies' take on platforms in their focus on media or the computational. What follows is something of a *historical typology*, a typology of platforms that acknowledges the historicity of the term, the key inflection points of its transformations, as well as the areas in which the term gains its definitional parameters. *Where* platform is theorized—in terms of geographical area but also the fields in which this theorization occurs—is as important here as *how* it is defined. As we saw with contents in the previous chapter, there is a politics and an effectivity to terminology—terms drive industry shifts as much as they reflect them, for which reason we must pay close attention to what terms meant at a given moment in time. These typologies lay the groundwork for an understanding of the platform that will gain analytical flesh over the course of the next chapters, particularly as we home in on platform practices in Japan. At a more fundamental level, the following also operates as a parsing of the mystifying multiplicity of meanings this word takes on.

There are two significant outcomes of platform discourse. First, it produces a distinction between a thing called a platform and a thing called contents. Second, this discourse produces a set of distinctions that allows us to better understand and isolate what a platform is, and what it does. The following will focus on the developments of the term in English-language literature as well as in Japanese, since they are parallel and mutually informing.

Here, I group these varying usages into three main types, creating a variation of a typology developed by management researchers Negoro Tatsuyuki and Ajiro Satoshi.

(1) Product-technology platforms

(2) Contents platforms

(3) Transactional or mediation platforms

This typology is partly a derivation from a series of distinctions that Negoro and Ajiro develop in their useful article "An Outlook of Platform Theory Research in Business Studies."[5] This roughly aligns with other overviews of platform literature from the management sector.

Yet there is something this management literature on platforms misses: the popular usage of the term in relation to social media sites, video uploading sites, and user-generated content sites. This is the body of objects that the term *contents platform* refers to. My modification of Negoro and Ajiro's typology makes room for the media studies and popular senses of the term *platform*. Finally, I should note that these typologies are not mutually exclusive categories—some existing platforms may fit into two or even three of these categories; they are analytic approaches as much as they are typologies of things.

This chapter and the next will build on their useful accounts and overviews of the platform. However, unlike other typologies of the term, I would like to combine typology with historical perspective on platform discourse—much as I did for contents in the previous chapter. Here I will chart its emergence in technological discourse around computers and computing, track its extension into product management, traverse the recent work in the subset of game studies known as "platform studies," follow its rise within media discourse and media studies, and locate, in the subsequent chapter, its theorization within a particular strain of managerial discourse that deals with the rise to prominence of "multisided platforms" or multisided markets. This chapter focuses on the first two types (product-technology platforms and contents platforms), including subdivisions of these, saving a full discussion of the third, transactional platforms, for the following chapter.

Hence, in what follows I hope to offer something like a synthetic account of the distinct (if sometimes adjacent) arenas in which platform discourse and its competing definitions appear, across geographical areas and a historically brief but fecund span of time. What results is not a "stacked" model of the platform—one layer built upon another, as platforms are often conceived.[6] Instead, this chapter offers an account that traces the conjunctions and adjacencies of platform discourse as well as the disconnections in their articulation. How a platform could be both an automobile chassis and a website where one clicks on "like" buttons interests me as much as creating some distinctions that allow us to see the difference between a Toyota car and a Facebook page. This chapter both narrativizes historical development and isolates distinct strains of platform theory. It traces the emergences and developments of the platform concept in different historical, discursive, and geographical contexts, and it also aims to disentangle the various usages of the term. In doing so, we may arrive at something of a synthesis that better enables us to understand what a platform is and does today, and how the very theory of the platform is indebted to disparate fields including automobile manufacture, computer history, management studies, and media analysis.

Product-Technology Platforms

Product-technology platforms refers back to the genealogy of computer hardware seen as platforms. This usage came into being in the late 1970s and early 1980s in the United States especially, rewriting an understanding of how a computer could become a hub for other peripherals like printers, and software built on top of its operating system. Game consoles and media technologies like the VHS or DVD players are also seen as platforms. Seemingly incidental but in fact crucial to this hardware genealogy are car platforms—standardized chassis that are the basis for multiple different car models. Software could also be considered a platform, and the Windows operating system is often analyzed as a platform within the computer industry. In all these cases, the technology platform tends to operate according to a layered model, wherein one layer forms the basis or support for the production of other things on the next layer. Questions of standardization and indeed monopoly also come into play, with economic benefits accruing to those companies whose technologies become the de facto standard (such as Microsoft's Windows operating system), accruing and deploying what David Singh Grewal dubs "network power."[7]

The usage of the term *platform* fluctuates between software, hardware, and later programming interface layers. This speaks directly to the ambiguity of the term, whereby it refers to different layers at different points in time—an observation made by tech writer Sasaki Toshinao, and supported by any analysis of the term's usage from within a historical perspective.[8] The continuity across these distinct usages lies in its *layered model*, with the platform acting as a base layer for another technology or media form.

Computers: Programmable Platforms

In both Japan and North America, it is fair to say the term *platform* is first associated with computers. In the computational context, the first sites where the term *platform* is used regularly is in relation to hardware. This is particularly the case in the English-language context where hardware-related uses overwhelmingly outweigh those for software in the late 1980s and early 1990s.

In Japan we find a similar situation. A 1998 book on Fujitsū's "cyber business" parses its four divisions; one is its "Information Processing Platform" division *(jōhō shori purattofōmu)* (the other three are "soft[ware] services," "communications infrastructure," and "electronic devices").[9] A 1994 annual white paper on computerization titled *Informatization White Paper (Jōhōka hakusho)* notes that the "aspect of computers as multimedia platforms has become increasingly important, given the astounding rise in cost performance of CPU and memory in recent years."[10] This definition links platform to the computer as a multimedia machine, a significant association that is particularly marked in the Japanese context. As we will see in what follows, it is the nature of the platform to act as a support for multiple media types or entities ("contents" or "soft"), and this link between platforms and multimedia is anything but a casual association.

In histories of computing, the first hardware platform is often said to be the IBM 360, first introduced in 1964, and released in 1965. While the use of the term *platform* in relation to computer hardware only appears to start in the early to mid-1980s, the IBM is generally historicized as the first computational platform.[11] In their 1997 article on the computer industry, and on the IBM 360 as platform, Timothy F. Bresnahan and Shane Greenstein identify "plug compatibility for hardware" and "technical standards for how the products worked together" as the "hallmarks of a *platform.*"[12] Here the platform designates a *shareable standard* that potentially benefits both the consumer and the producer of the standard. In this regard, part of the benefits offered by the platform is interchangeability, or compatibility: "Platforms have interchangeable components, so many buyers can share the benefits of the same technical advance. Interchangeable components also permit buyers to use the same platform over time, avoiding losses on long-lived software or training investments."[13] Platform analysis such as that around the IBM 360 aims to garner insight into why the concentration of platforms occurs—why, that is, Windows and Intel as "Wintel," or IBM before them, become monopoly powers. The same conclusions could be brought to bear on the market dominance of Android and iOS in the mobile environment today. Indeed, the very premise of Bresnahan and Greenstein's analysis is an even wider-ranging and prescient argument about the need to shift from the firm as the unit of analysis to the platform: "Much of our analysis of the forces for stability in industry structure drew on well established competition and industry structure theory. To make it fit the computer industry, we changed a focus on the firm to a focus on the platform."[14] This shift from the firm to the platform as the unit of analysis has become the common sense within management literature today, and is the premise of managerial literature around platforms we will engage with in the subsequent chapter.

For a close—and indeed polemical—engagement with the concept of platform, it is worth looking at Netscape Navigator browser inventor and venture capitalist Marc Andreessen's definition of the term. In his discussion of "internet platform" in particular, Andreessen first notes the discussion of the platform extends back to the beginning of computing: "We have a long and proud history of this concept and this definition in the computer industry stretching all the way back to the 1950's and the original mainframe operating systems, continuing through the personal computer age and now into the Internet era."[15] In fact, a thorough examination of the Factiva database parsed semantically indicates that the use of the term *platform* in relation to computing only begins to be used in the mid-1980s, increasing in frequency into the 1990s. Until this time, platform was used predominantly to refer to oil rigs, political platforms, as well as other metaphorical (or materially specific: platform shoes) but noncomputational uses.

Putting historical accuracy aside, in the September 2007 blog post cited above, Andreessen works to narrow the definition of platform, with specific reference to an "Internet platform." There he writes: "'Platform' is turning into a central theme of our industry and one that a lot of people want to think about and talk about. However, *the concept of 'platform' is also the focus of a swirling vortex of confusion*—lots of platform-related concepts, many of them highly technical, bleeding together." He continues: "I think this confusion is due in part to the term 'platform' being overloaded and being used to mean many different things, and in part because there truly are a lot of moving parts at play that intersect in fascinating but complex ways." As the base element of his definition, Andreessen, quoting from one of his own June 2007 blog posts, explains:

A "platform" is a system that can be programmed and therefore customized by outside developers—users—and in that way, adapted to countless needs and niches that the platform's original developers could not have possibly contemplated, much less had time to accommodate.[16]

In the September 2007 post, he continues:

So, if you're thinking about computing on the Internet, whenever anyone uses the word "platform," ask: *"Can it be programmed?"* Specifically, with software code provided by the user? If not, it's not a platform, and you can safely ignore whoever's talking—which means you can safely ignore 80%+ of the people in the world today who are using the term "platform" and don't know what it means.

Key here is Andreessen's refusal to cede the ground of platforms to what might have been the most prominently called platform of his day—YouTube, or social media sites in general. I will return to these below, but for the moment it is simply worth noting that this is emphatically *not* what he is discussing (unless in the narrow sense of Facebook's then-recent opening to outside developer application programming interfaces [or "plug-in API"]—which allows outside developers to use Facebook as a platform for their particular services). Summarizing his earlier post, Andreessen posits in the September 2007 blog that "the key term in the definition of platform is 'programmed.' If you can program it, then it's a platform. If you can't, then it's not."

In his earlier post on the Facebook API, Andreessen offers the following distinction, which I will quote at length.

> Definitionally, a "platform" is a system that can be reprogrammed and therefore customized by outside developers—users—and in that way, adapted to countless needs and niches that the platform's original developers could not have possibly contemplated, much less had time to accommodate.
>
> In contrast, an "application" is a system that cannot be reprogrammed by outside developers. It is a closed environment that does whatever its original developers intended it to do, and nothing more.

The example Andreessen offers of this distinction is pervasive in computer history lore: the distinction between *a single application device* and *the programmable personal computer*. Andreessen writes in this same post: "The classic example of an application being vanquished by a platform was the Wang word processor versus Microsoft DOS-based personal computers."

This distinction between a single-purpose device and the multipurpose computer is instructive. The computer is a platform, whereas the device is merely a single-application machine. This accords well with the distinction internet scholar Jonathan Zittrain makes in *The Future of the Internet and How to Stop It,* where he distinguishes what he calls "tethered appliances" from the "generative computer."[17] For the former "tethered appliance" he is thinking of the iPhone in particular, which despite having the potential to be a computing machine was configured in such a way to make it a closed, nonprogrammable device (note that his book was written before the iPhone was "opened" to outside developers with the advent of the App Store and third-party developer support—essentially responding to criticisms like Zittrain's or, perhaps more cynically, reflecting on the reason for the historical dominance of the Windows operating system over the Mac, and recognizing the benefits of the network effects that tend to accrue to more open systems). For Zittrain, the iPhone of 2007 was more like a toaster, or Andreessen's Wang word processor, than a multipurpose computer. The iPhone was for Zittrain a glorified single-purpose appliance. The PC computer, on the other hand, was "a platform rather than a fully finished edifice, one open to a set of offerings from anyone who wanted to code for it."[18]

In this context, a platform means a fundamentally *open* and hence transformable technology; it is a device whose end result is not determined from the outset (unlike, say, the toaster that can only really toast bread) but rather can be modified by the user in a presumably fundamental manner, such that the apparatus can perform radically different functions—from writing a dissertation to editing a film. This reflects of course the classic understanding of the computer as metamedium, or, as we saw Fujitsū term it, a multimedia platform.[19] Computer as platform or internet as platform is in this sense fundamental extensions of the West Coast tinkerer ethos that is manifested as computer: a programmable and alterable entity—a basis upon which some yet unknown or unrealized function can operate. This conceptual framework is at the very core of product technology definitions of the platform, positing it as a foundational layer built to accommodate subsequent and as-yet-undetermined layered installments. It is open to creativity and to whim; it is an *open platform upon which further platform layers could be built.*

Conversely, and rather interestingly, the closed has also figured as an essential element of the platform in literature on the subject; the distinctions between open and closed are also operative and important in discussions around particular internet subsystems in general. Early modes of accessing the internet such as America Online (AOL) were exemplars of "walled-garden" approaches to the internet, as opposed to the more open World Wide Web model. AOL gave users easy access to a limited scope of web pages and services, all contained within AOL's walled garden. This closed model continues to operate in the increasing enclosure of the web into fiefdoms, represented first and foremost by Facebook. Facebook, in turn, is one of the preeminent actors within the platformization of the internet, as Anne Helmond so convincingly shows (a complex process wherein the tentacles of the walled garden reach far outside the garden, tunneling into the open web through APIs and Like buttons, and bringing traffic and data back to Facebook's walled-off fiefdom).[20] But the very figuration of Facebook in relation to the platform suggests that closed systems are as much platforms as open ones.

So the open versus the closed is far from settled in debates around the platform, and often a platform is best described as a particular balance between being open and closed. This degree of aperture or closure also coincides with geolocking or other forms of geographical closure that should not be ignored.[21] That said, the closed model is in fact an inspiration for the Japanese cell phone internet service, which was conceived according to the relatively closed AOL model, even as it attracted third-party content providers and later served as the model of a particular mode of content delivery that presumes walled gardens. In short, relatively closed models of the platform—or hybrid models of open and closed—are increasingly the norm.

Console Wars: The Game Platform

Zittrain's equation of open with the personal computer and closed with tethered devices also excludes a major category of "platforms" as they are usually called: video game consoles. The video game console is generally a closed computing system that nonetheless goes by the name *platform*. Yet a similar multiple-layered definitional sense structures discussions of the game console. To be sure, the degree of programmability varies with the console. However, one of the underlying elements of game consoles is their closure—they are designed *not* to be programmed by just anyone but rather to play the very specific games that have been designed for them. In Zittrain and Andreessen's formulation, insofar as they cannot be programmed by the user, they are closer to appliances. And yet this appearance of being nonprogrammable is counterbalanced by their programmability via other software programs, in a word, games. Video game software is what activates the console as such. It is also what turns the console into what we would call a platform today.

The *Merriam-Webster* dictionary defines the console in the following manner: "an electronic system that connects to a display (such as a television set) and is used primarily to play video games."[22] The game console is at the heart of Ian Bogost and Nick Montfort's definition of the platform for their Platform Studies book series at MIT Press, one of the most coherent and important initiatives within media studies to put the platform on the map of media studies research. Bogost and Montfort propose what they frame as a strategically limited version of platform studies, one they claim to have "introduced as a concept" in 2007. This is a rather important development, in large part because of the overweening influence it has had within film and media studies as a model for the study of the "platform." For Bogost and Montfort, "'Platform studies' is a new focus for the study of digital media, a set of approaches which investigate the underlying computer systems that support creative work."[23] Despite its focus on game consoles, their series is not limited to the study of video game platforms, inviting contributions on *"all computing platforms on which interesting creative work has been done,"* including software platforms. They write: "A platform is a computing system of any sort upon which further computing development can be done. It can be implemented entirely in hardware, entirely in software (which runs on any of several hardware platforms), or in some combination of the two." They suggest that the Atari VCS is an example of a platform that is entirely hardware, while the Java programming language (meant to run on all operating systems regardless of hardware specificity—its multiplatform credo being "write once, run anywhere") is an example of a software platform.[24]

In their inaugural contribution to the Platform Studies book series, *Racing the Beam: The Atari Video Computer System,* Bogost and Montfort tend to use the term to mean *game platform,* or a kind of *computational hardware assemblage*—although they acknowledge platforms are, at the first level, abstractions or standards. In *Racing the Beam,* Bogost and Montfort focus on the Atari 2600 and explain: "The Atari Video Computer System . . . is a well-defined example of a platform. A platform in its purest form is an abstraction, a particular standard or specification before any particular implementation of it. To be used by people and to take part in our culture directly, a platform must take material form, as the Atari VCS certainly did."[25] In their usage, the platform is the set of material constraints that suggests the contours of any creative practice; in their view, this practice can only be understood through and by a literacy in the computational hardware (and hence limitations) that subtend it. Such an understanding will allow them and the authors in their series to "reverse engineer" aesthetic decisions as technical ones. In some sense, what we end up with is a technically sensitive aesthetics of the console and its affordances.

What is most useful about the "platform studies" initiative is that Bogost and Montfort develop it as a methodology, a mode of studying and approaching platforms.[26] Yet the use of the term *platform* in platform studies in relation to predominantly *game console hardware* is overdetermined by the prevalence of *platform* as an alternative to *console* within the game industry and popular parlance. (Bogost and Montfort include a caveat that their series is not meant to solely focus on game platforms; moreover, as the series has progressed, they have also embraced work that draws on the economic or transactional definition of platforms that informs this book, and that increasingly embraces the cultural dimensions of the platform as an equally significant driver of platform building as the hardware they insisted upon at the beginning of the series.[27]) In their analysis of the platform studies endeavor, Thomas Apperley and Jussi Parikka write that "platforms are not recalled and rediscovered through platform studies, rather *in the process of 'doing' platform studies, a uniform platform is produced.*"[28] Yet the reverse is also true: the very impetus of doing platform studies comes from our preexisting disposition to calling game consoles platforms; the uniform object of platform is produced as much discursively within gaming culture as through the process of doing platform studies. It is on the basis of this existing, recursive tie between game consoles and platforms that Bogost and Montfort can justify their use of the term and in turn criticize its use in other contexts (such as Tarleton Gillespie's discussion of YouTube as platform). The use of *platform* in relation to YouTube gives rise to the nexus of "confusion" to which both Andreessen and Bogost and Montfort (who cite Andreessen approvingly) respond—and seek to differentiate themselves.

What this formulation of the game platform occludes, of course, is the all-important element of the business model that the platform implies. As Gillespie properly argues, "'Platforms' are 'platforms' not necessarily because they allow code to be written or run, but because they afford an opportunity to communicate, interact or sell."[29] The ability to sell and the business model behind it—and not simply the code—are what make a platform such. Nintendo, for instance, is famous for having perfected the market relationship between console and games. By pricing the console at cost and games at a premium, Nintendo created a virtuous circle that took advantage of positive network effects: the more game consoles sold, the more consumers there would be for its games. Nintendo carefully managed these network effects through multiple strategies: by strictly limiting which companies could make games; by controlling these games for quality; by including a lockout chip in their consoles that prevented unauthorized game making; by locking consoles and games to specific regions or territories; and by strictly rationing the production of chips, at the same time as they exerted extraordinary control over the game producers, requiring an expensive outlay for the purchase of these chips.[30]

Hence Nintendo is as famous for its game consoles as it is for the business model that accompanies them, and which impacts later "business platforms" from the PlayStation to i-mode, to the Apple Store and Google Play Store. Notably, the ingraining of its business model into its hardware via mechanisms like the lockout chip effectively engineers and reinforces the distinction between hardware console and software package, between platform and contents, engineering an early form of digital rights management, as Nathan Altice notes in *I Am Error.*[31]

The takeaway here is that console developers "solved" the problem of how to limit the circulation of content on computing platforms, and hence would become one of the models for the separation of content from platform in cell phones and later smartphones, reinforcing the *layered model of the product-technology platform,* and the distinction between platform and contents this layered model supports. Before I turn to the integration of platform and contents in the "contents platform" model, allow me to briefly address the crucial place of studies of automobile manufacture and Toyotism in particular in the development of platform theory in toto.

Product Platforms: Excursus on Automobiles

This figuration and use of the term *platform* (as layered product-technology) extends to automobile production at the end of the 1980s and early 1990s. While initially it may seem like it is outside the current semantic purview of platforms, automobile platforms in fact intersect with current platform discourse in key ways, not the least being as logistical solutions to supply chain management, as well as being base elements or standards that allow for further variation—not so dissimilar from the hardware platform to software relationship we have seen above. Perhaps even more importantly, the theorization of automobile production as platform involves (directly and indirectly) some of the key people who will later pioneer a full-fledged body of platform theory in North America—namely Anabelle Gawer and Michael A. Cusumano. Despite the term's natural home in the computational, it is automobile production that is the first fledgling site of the generalization of the term *platform* from the realm of computation to a reconceptualization of products and production techniques more broadly. It also figures into the eventual description of capitalism itself as being platform in nature. Critics such as Nick Srnicek note that the transition from Fordist to post-Fordist or Toyotist models of manufacture is one of the industrial conditions of the rise of platform capitalism.[32] As we will see in what follows, studies of Toyotism were also critical to the very development of platform theory, at this very moment of transition in the configuration of labor and capital. Hence we can locate Japan at the epicenter of the American development of platform literature—albeit in an attempt to make sense of automobile manufacturing techniques pioneered in Japan.

The use of the term *platform* in the automobile industries is one of the first sites where we see an overt combination of hardware production and business strategy. This usage implies a certain transformation of assembly lines with an eye to building multiple models of cars from a single "base," or standard, and coincides with the optimization of automobile production under the shadow of the "Toyotist" system that shook up the global automobile industry during the 1970s and 1980s. Despite occasional uses in the early 1970s, we can date the popularization of the term *platform* in the automobile industry to the late 1970s, with its usage increasing by around

1978–79, becoming the go-to term by the 1980s.[33] Platform denotes a standardization of the base element of the car—the ground, or foundation—upon which different bodies could be placed. One article from 1979 glosses the platform as "the base-shell on which a car is built."[34] A 1980 report to the U.S. government describes the "now almost universal acceptance of the platform strategy (one basic car design that can be stretched or shortened without complete retooling of all phases of the production process) to cut production costs."[35]

The concept of a base-level standard that reduces the cost of car models or iterations built upon it precedes the designation "platform." Some precedent terms seem to be "model," "base model," "base-shell," or "body shell"—terms that persist even with the rise of "platform." For instance, James Womack, Daniel Jones, and Daniel Roos's classic 1990 business book examining Toyotist versus Fordist models of production, *The Machine That Changed the World,* puts it in the following manner: "Automobile development projects can vary greatly in the size and complexity of the vehicles, the number of different body styles spun off a base model (or 'platform,' in car talk)."[36]

While the lineage of automobile platforms might initially be considered far outside the general semantic and theoretical parameters of the term, it is in fact particularly pertinent to platform discourse for two reasons I would like to highlight here. First, insofar as the development of this car platform is part of the optimization of car production, it is profoundly involved in a rethinking and rejigging of supply chain management that is in turn a key element of logistical thinking. Logistics and the circulatory logic it invokes—as well as the drive toward optimization upon which to build other businesses and economic transactions—is deeply imbricated with the rise of the platform economy and the shift to platform capitalism.[37]

Second, the development of the "platform concept" within Japanese management literature in the 1990s (the subject of the next chapter) appears to be cognizant of the usage of the term *platform* within the automobile sector and references developments within the study of the automobile industry as one parallel type of platform study. In this regard, one of the earlier formal definitions of the platform in Japanese managerial literature would be found in the work of Nobeoka Kentarō, an automobile industry analyst who studied within the MIT International Motor Vehicle Program (IMVP). Much of the empirical and theoretical work on Toyotism and "lean production" was first developed there in the 1980s as a response to transformations in the auto industry. In his 1996 Japanese book on "post-lean" management of product development, *Multiproject Strategies,* Nobeoka defines the platform as the "core technology of an automobile" that "determines the general outlines of an automobile's basic structure (architecture), and is composed most centrally of parts like the floor pan and suspension."[38]

From its usage within the automobile sector, the term *platform* expanded outward to cover other "products," becoming a subgenre of literature on what became known as the "product platform." Steven C. Wheelwright and Kim B. Clark are generally credited with the expansion of the term from the auto sector outward, and they draw the automobile industry's framing of the term as both a system of parts and a logic of manufacture to describe a process within the design of a product.[39] Notably, Wheelwright and Clark refer to automobiles as one of their first examples, writing: "Honda's 1990 Accord line is an example of a new platform in the auto industry: Honda introduced a number of manufacturing process and product changes but no fundamentally new technologies."[40] They continue with a reference to computers before expanding to a wide range of products, including Tide detergent.[41]

Fernando F. Suarez and Michael A. Cusumano, in their extensive overview of the "product platforms" literature, similarly note the importance of the automobile usage of the term to "product platforms" work in general:

> The word "platform" became commonly used in management within studies of product development, particularly in the automobile industry. . . . In this literature, a platform was considered to be a set of common components and a general design or architectural "blueprint" that supported product variations and extensions through part substitution and part extension. The auto industry was probably one of the first to adopt a platform strategy, by which products that are apparently very different (such as the Toyota Camry and the Toyota Highlander SUV, in Toyota's current lineup) actually share a common underbody and many other components.[42]

As Suarez and Cusumano explain, the automotive industry was the first to start employing the term *platform*. Cusumano himself was a major player in MIT's International Motor Vehicle Program, as well as Nobeoka's PhD supervisor and coauthor on a number of important articles.

More crucially still, Cusumano was the coauthor of one of the most impactful early management books on platforms in the early twenty-first century: *Platform Leadership: How Intel, Microsoft, and Cisco Drive Industry Innovation*. Cowritten with Anabelle Gawer, and largely based on her PhD dissertation on Intel, with additional chapters coauthored together, Gawer and Cusumano's *Platform Leadership* put the concept of platform into even wider circulation, and continues to be a touchstone today. Common to both this work and its automobile platform precedent is the supposition that the platform is the base on which to build products in general, or product families. Institutionally it is significant that platform theory emerges from the IMVP, a program designed to stimulate the study of Japanese automobile manufacture, especially during the 1980s. Hence we find Japanese hardware (in this case automobile production) to be at the discursive nexus of platform theory in North America.

This layered structure is common to all *product-technology platforms* we have seen in the first part of this chapter: computer hardware, software, internet-based APIs, video game consoles, automobiles, and computer chips. This base-level standard or technology, on top of which comes a next layer application—*contents* being the best word for this x factor—is at the core of the platform concept. Picking up where we left off in the previous section on debates between hardware-centric and web-based utilizations of the platform concept, I would like to turn to one of the most popular usages of the term in recent years: Web 2.0 sites as platforms.

Contents Platforms

The end of the first decade of the twenty-first century saw a distinct shift in the usage of the term *platform,* from something predominantly designating computer hardware or software or game consoles—in Andreessen's as well as Bogost and Montfort's usage of the term—to a term used to refer to social media networks and video streaming sites, or rather, what I am calling *contents platforms*. The most popular usage of the term comes in reference to content-sharing websites like Facebook and YouTube, particularly within media studies, where the two have become the most frequent referents of "platform."

In Tarleton Gillespie's rigorous analysis of the term in "The Politics of 'Platforms,'" corporations like YouTube make strategic use of *platform* to claim immunity from prosecution. Deploying the term becomes a strategy for corporate positioning in an era of battles over copyright infringement in the United States (particularly during the moment of YouTube's emergence). Rejecting hardware-centric uses of the term, Gillespie suggests that for better or worse, "in the discourse of the digital industries, the term 'platform' has already been loosened from its strict computational meaning. Through the boom and bust of investment (of both capital and enthusiasm), 'platform' could suggest a lot while saying very little."[43] The term is instead mobilized to immunize websites that rely on user-generated content (UGC) from copyright-related prosecution. (In more recent years, including in the immediate aftermath of the 2016 U.S. presidential election, this included the attempted defense by Facebook that it should have no role in mediating or censoring fake news stories—a stance it eventually backtracked on, but which shows the continuing appeal to the "we're just a platform" defense.) The term's other connotations have the added benefit of making an appeal to the general public's sense of the ownership of these otherwise very proprietary sites. That is to say, the popular sense of platform as something we all share, that we all have a stake in, hides the proprietary nature of sites like YouTube and Facebook. These sites are based on user-generated content, but users do not have control over them. Quite the opposite, these sites function and make money by gathering user data. The chance to generate content is, as Mark Andrejevic and others have argued, a mere enticement to users, and a tool to get them to hand over their data and personal information. It is this personal data that these sites are really after.[44] As José van Dijck has put it: "For many platform owners, content is just another word for data; they are particularly interested in the quantity of data streams that flow through their channels, which they can treat as aggregated and computational resources."[45] Data, *The Economist* agrees, is now the "world's most valuable resource"—the twenty-first century's equivalent of oil.[46]

The usage of platform in relation to UGC sites can, moreover, be dated to around 2006. Gillespie opens his article with several quotes from the 2006 announcement of Google's acquisition of YouTube. Gillespie continues: "A few months later [in 2007], YouTube made a slight change to the paragraph it uses to describe its service in press releases. This 'website,' 'company,' 'service,' 'forum' and 'community' was now also a 'distribution platform for original content creators and advertisers large and small.'"[47]

What of the case of Japan? An examination of two annual white papers, the *Jōhōka hakusho* (Informatization white paper) and the *Jōhō media hakusho* (Informational media white paper) published in Japan between 1993 and 2011 reveal a similar shift in vocabulary; "digital intermediaries" or "content intermediaries" (as Gillespie describes them) known as "video contribution sites" *(dōga tōkō saito)*, or simply "sites" *(saito)*, began to be called "platforms," ushering the term into a new era of popularity and common usage.[48] If in the United States this usage begins around 2006, in Japan it seems to take root several years later, around 2009–10. The term *platform* was previously used only in relation to hardware or chipsets like Intel during the 1990s; the computer as "platform for multimedia" in the white papers finds a new life in relation to social media sites and video sites in particular around the early 2010s.[49] Examining the annual *Informational Media White Paper* reveals 2010 as the year *platform*—specifically, *video delivery platform (dōga haishin purattofōmu)*—replaces earlier terms when referring to sites like YouTube or Japan's Niconico Video.[50] The following year's report finds the consolidation of what were previously referred to as "main sites," or websites under the category "platform," in a graph that simplifies the earlier, more complex graph with one that emphasizes three main sections: contents, platforms, users.[51]

Upon first glance, then, it would seem that the term *platform* in relation to UGC sites such as YouTube comes into use in Japan after its introduction in the U.S. context, in the years following YouTube's purchase by Google. Two caveats complicate this picture, however. First, in the mobile phone section of the *Informational Media White Paper,* the term *platform* was deployed to discuss the internet services of mobile phone providers as of 2008. Indeed, the term *platform* was already in use in the very early twenty-first century by mobile industry people such as Natsuno Takeshi, in reference to i-mode. I-mode, an AOL-like service for phones, which I will return to as a key platform in chapter 4, is one kind of "platform" listed in this graph, which is divided into four sections: Contents, Platforms, Networks (infrastructure/services), and Terminals.[52] This shows that *platform* is already being deployed to name a mediating structure for the diffusion or intermediation of contents. Second, another usage of platform comes in the *Informatization White Paper*'s discussion of cable satellite (CS) broadcasting in Japan. The term *platform* first appears in passing in the 2001 edition of *Informatization White Paper,* with reference to SKY PerfecTV.[53] Readers must wait until the following year's report to be provided with a definition, and explanation, of what exactly the authors meant by this term.

> The particularity of CS broadcasting is a concept which has not existed in broadcasting until now: the platform. Regular broadcasting stations operate by producing and financing their own contents, and then broadcasting these contents from their own facilities. By contrast, CS broadcasting is separated into the business operators who commission the production and financing of the contents; the business operators commissioned to operate the broadcasting system; and the business operators (the platform) who do the billing and marketing activities.[54]

This account of cable satellite broadcasting as a platform at once intersects with existing discussions of the platform as an intermediary in business literature—which we will come to in the following chapter—and coincide with parallel considerations of mobile internet services such as i-mode as platforms. It also suggests another distinct area in which platform discourse takes root: broadcasting, an area on which Thomas Lamarre has done important work.[55]

These caveats aside, the emphasis on delivery or intermediary suggests something specific about these "new" kinds of platforms: while they sometimes depend on the layered sense that we find in the product-technology platforms examined in the first part of this chapter, they also play on a sense of middling or intermediary function of distribution that is not present in the former, marking a clear shift away from the hardware uses that we see during the 1990s and the beginning of the following decade.

This sense of platforms as a distribution channel for contents is clearly articulated by Kawakami Nobuo, CEO of Kadokawa and founder of the Niconico video sharing platform. He offers a useful reflection on both the term and its object in his 2015 book, *The Future of the Internet That Even Mr. Suzuki Can Understand.*[56] The "Mr. Suzuki" of the title is none other than the famed anime producer Suzuki Toshio, the financial force behind Studio Ghibli, and directors Takahata Isao and Miyazaki Hayao's long-time partner. He is also a person with whom Kawakami apprenticed, and to whom the book is ostensibly directed, an explanation of the present and future of the internet and net culture addressed to a general audience (and a curious Suzuki).

Kawakami contributes to discussions of platforms by specifying a very particular kind that he focuses on: the "platform for contents" or, as I will translate it here, the "contents platform" *(kontentsu no purattofōmu).* This likely builds on an existing usage that appears, for instance, in tech writer Sasaki Toshinao's 2007 book, *Netto mirai chizu* (Net future map), where he describes telecoms that provide content (such as i-mode) as "contents platforms."[57] Kawakami writes: "The word platform itself is used here and there in the world depending on the issue in question, with subtly different meanings, so I will explain the meaning of the term in the context of a platform for contents."[58] The contents platform is "the framework that circulates content," writes Kawakami, calling this framework a "format."[59] He adds:

> Platforms for circulating contents have the following roles to play:
> - Offering a business model
> - Offering a user base
> - Offering a means of promotion
> - Defining the framework (format) of contents
> - Controlling the quality of contents.[60]

As he outlines here, the platform does not only distribute contents; it also controls contents and functions as a mediator between users and advertisers—not unlike the way television stations used their programs to gather eyeballs and sell them to advertisers in the era of broadcast television.

This aspect of mediation will come out more clearly in the work on platforms I discuss in the next chapter. Insofar as platforms are defined by their goal of hosting and circulating contents, they are also, in Kawakami's account, in a position of superiority relative to these contents. His advice to contents makers: "Ideally, the contents side should own a platform"—here clearly thinking of both Ghibli's future as well as Dwango's own recent merger with contents-maker Kadokawa.[61] This is also a lesson that comes from Dwango's previous dependence on mobile platforms (as a contents producer, at the time), and the existential threat the shift from i-mode to Android/iOS posed for the company. As we will see later in this book, the sudden loss of income that attended this shift led Dwango to rethink its business model, create Niconico Video, and ultimately rethink the relationship between platform owners and contents producers. I will return to issues of monopoly and platforms in my discussion of platform imperialism in chapter 5, but at this point it is worth signaling the power imbalances that Kawakami draws our attention to—power imbalances between platform owners and the contents they distribute or ask users to create. In Kawakami's telling, it is the platforms that will always be king, since they set the rules, conditions, and most importantly the prices for the distribution of contents. What you can see or not see, what you can charge for or not, what percentage cut the platform takes—all these conditions are set by platform owners, putting the latter firmly in the driver's seat.

Relational Platforms

Contents platforms tend to be framed in terms of the collective, the mutually owned or shared—a vision of users generating and sharing content for other users, without the intermediaries of big media companies. Of course, as Gillespie and others have pointed out, this formulation is a fiction that serves the interests of equally large media companies that come from the tech world rather than the traditional media of television, newspapers, and publishers. That publishers and other contents producers are also reconceptualizing the platform is, interestingly, visible in Kawakami's own account—as he moves the term more explicitly toward the nontechnological sense of platforms as mediation or mediators that we find in business literature, and particularly in the writings of business leaders who see themselves as being in the platform business. These eschew the false pretense of the focus on the people as the mediators and focus attention instead on the function of mediation, recalling José van Dijck's suggestion that we consider platforms through the Latourian framework of "*mediator* rather than intermediary."[62]

For instance, Kadokawa chairman Kadokawa Tsuguhiko offers a simple yet provocative definition of platform: it is the place where money, people, and commodities meet.[63] This in turn has echoes in Google executives Eric Schmidt and Jonathan Rosenberg's definition of the term in their *How Google Works,* where they write: "A platform is, fundamentally, a set of products and services that bring together groups of users and providers to form multisided markets."[64] At first glance, there is something novel and even counterintuitive about this formulation of platform. It is no longer something necessarily digital *or* medial—it doesn't necessarily refer to something that is either a media form or computational in the widest sense. For Kadokawa, a publisher at the core, bookstores are a key example of platforms: gathering places that bring money, people, and media together. Schmitt and Rosenberg similarly acknowledge that "platforms are increasingly (if not exclusively) technology based"[65]—leaving room for nontechnological understandings of the platform. In *Platform Capitalism,* Nick Srnicek offers a broad, initial definition that nonetheless hews closely to this way of thinking about platforms: "At the most general level, platforms are digital infrastructures that enable two or more groups to interact. They therefore position themselves as intermediaries that bring together different users: customers, advertisers, service providers, producers, suppliers, and even physical objects."[66] Srnicek emphasizes the digital nature of platforms as intermediaries; at the same time, if one removes the digital from his definition one can find in his stance a position similar to that of Kadokawa—in short, a site of commodified encounter.

Platforms in all these accounts are figured as *structures of relation,* or the formalization of relations in varying degree of technological means. As Kadokawa and Schmitt and Rosenberg emphasize, these platforms are not necessarily digital.[67] So what does it mean to suggest a non-medium-specific definition of the platform, and from where does this definition come? To answer this question, we have to turn to the lineage of platform research within management studies in Japan and within microeconomics in the English-language context that develops the model of "transactional or mediation platforms," a nontechnological conception of the platform-as-mediator.

Moving from the technological discourse on platforms, to the functionalization of the layered model within technological platform discourse, to an expanded conception of the product-technology platform that includes automobiles, to, finally, the user-generated contents-platform hubs of the last section, this chapter has covered the historical development of two types of platform discourse and objects in the United States and Japan. In so doing it has also addressed some of the critical issues in debates around the platform, from the open versus closed narratives, to the way studies of Toyotist automobile production flow out to inform other product-related uses of the term. Lurking behind all of these uses is another, even more crucial definition of the term—the platform as economic intermediary, to which I turn in the next chapter.

CHAPTER THREE

Transactional Platform Theory

A Japanese Genesis

The keyword *platform* can encompass many things. From computer or game console hardware, to software systems (like Windows or Java), which I call *product-technology platforms*; to collective or user-contributing sites like YouTube and Niconico Video, which I call *contents platforms*; to, finally, a "place where money, people, and commodities meet." This third definition represents what I term (in a variation on Negoro and Ajiro's terminology) *transactional or mediation platforms,* and what falls under the "industrial economic" strain of platform theory, which is the main subject of this chapter.[1] In fact, this is arguably one of the only bodies of work that offers a *theory of the platform* as such—rather than simply a definition of the platform as an object (whether computer, automobile, or game console). For this reason, the Anglophone and Japanese bodies of work on the platform are central to coming to terms with platform theory in toto. Despite its importance, though, this work is only starting to be treated within media studies.[2] Part of the work of this chapter is to show how this "transactional or mediation platform" represents a distinct and crucially important lineage within the short history of the platform concept. It is ultimately this usage more than any other that defines platforms today, and shapes the platform economy as we know it.

There is a vibrant body of literature in English and in Japanese on transactional or mediation platforms. As is often the case in non-Anglophone, non-Western countries, the Japanese writers on the topic are deeply aware of the English work, but the opposite is not the case—an all-too-common imbalance in the geopolitics of knowledge production. More interesting for me here, however, is that the Japanese-language work precedes the North American theory of mediation platforms in the business context, anticipating by as much as a decade many presumptions at the heart of the platform literature in English penned by French and American economists. Let me begin with an account of the latter, sketching both the substance of the questions asked within it and the timeline of its development. What is particularly important in this lineage is to see the moment when platform ceases to be tied to a material substrate and becomes something of a relational entity, or intermediary as such. (This process has parallels with the dematerialization of media that gives rise to contents discourse, as discussed in chapter 1.)

The aim, then, is to do for the platform what John Guillory has done for the now ubiquitous concept of "media": to trace the key moments in its development and chart the *genesis of the platform concept*.[3] While I cannot do justice to the historical span found in Guillory's impressive account, I will do something his intervention does not do: question the European origins of the media concept, in this case by examining the platform concept outside the presumed French origins of the term, or the American elaboration of economic platform theory. Indeed, as we will see, much of platform discourse as it exists today has some debt to Japan—whether technological or discursive. Japanese platform theory of the 1990s precedes the French and American platform theory in management and economics that is more familiar within the English-language context. The genesis of the platform concept occurs, at least in part, in Japan. In what follows I will closely examine both bodies of theory, as they are each key to platform practice. In so doing I take quite seriously work on the edges of management theory, a body of work that is similar to what Alan Liu has called "the immensely influential and best-selling works of fiction-blended-with-realism—let us loosely call them 'novels'—by the Victorian sages of our time: 'management gurus.'"[4] Platform theory forms a body of work that we might dub *executive theory*—a theory of platforms written for and read by business executives, aspiring entrepreneurs, managers, and so on (and thus has overlaps with what John Micklethwait and Adrian

Wooldridge call the "management theory industry"[5]). The practical impact of this executive theory will become apparent in the subsequent chapters when I address the actual deployment of this platform model and its formative influence on the mobile internet. Yet, as crucial to the development of platforms as this theory is, it is equally significant for the ways it expands the role of management itself, justifying the increased control over a platform's users or consumers. While prior theories and practices of management limited themselves to the internal management of a company (creating efficiencies in work, organization, and so forth), this particular branch of management takes the entire social field as its site of managerial control. And this is the true reason platforms not only become ubiquitous but also extend their control over entire societies, leading to platform capitalism as such.

Platform Theory in the United States and France

A good place to start our exploration of platform theory is Eric Schmidt and Jonathan Rosenberg's definition of the term *platform* from *How Google Works,* quoted in the previous chapter, as it draws directly on the body of work responsible for advancing platform theory. As they write: "A platform is, fundamentally, a set of products and services that bring together groups of users and providers to form multisided markets."[6]

As they indicate in their definition and the note that follows, their account of the platform is indebted to work that comes out of a subfield of microeconomics, or industrial economics, concerned with markets in which there are two or more "sides"—known as two-sided or multisided markets. The objects that enable these markets are platforms, hence the most common appellation for these are "multisided platforms," abbreviated as MSP.[7] This work attempts to account for scenarios in which a given company deals with two or more markets simultaneously. The most prominent examples in this regard are video game console makers (which have consumers of games on the one side, game developers on the other, and consoles mediating the two), software companies like Microsoft (which have end users of Windows, software developers, and hardware manufacturers as their three markets), and credit card companies (which have card users on the one hand, and merchants on the other)—to name a few.

The principal variable in markets like these is how much to charge one side, and whether a company should subsidize another (usually one side is subsidized, and the other subject to charges). These markets are ultimately supported by what is known as "network effects." "Same side network effects" (also called "direct network effects") dictate that, for instance, the more users there are of a particular social media site, the more value the site has and the more other people will want to become users. The telephone works in the same way. A single telephone is worth nothing—who would one call?—but the more people who have telephones, the more the telephone is valued. "Cross side network effects" ("indirect network effects") dictate that an augmentation or subtraction of users on one side will impact the actions of the other side. For instance, the more users of a social media site there are, the more software companies want to develop additional applications for the social media site in question.[8] That *thing* that mediates between the two sides is first referred to as a "firm" but between 2000 and 2003 comes to be called a "platform."

Among the main sets of writers at the heart of the two-sided markets literature are Jean-Charles Rochet and Jean Tirole, based in France. Their "discovery" of a particular kind of business that they term *platform* in a research paper first presented around 2000–2001 is credited with kicking off the literature around multisided markets by David S. Evans and Richard Schmalensee.[9] U.S.-based academics such as Evans and Schmalensee, as well as Geoffrey G. Parker and Marshall W. Van Alstyne, were equally key in the development of platform literature in the early twenty-first century. In Parker and Van Alstyne's version of events, they and Rochet and Tirole were working simultaneously along the same sets of problems—dealing with two or more–sided markets. Indeed, Parker and Van Alstyne's article from 2000, "Information Complements, Substitutes, and Strategic Product Design," goes to the heart of the platform problem—"how firms can increase profits by giving away free product"—which they locate temporally in the "information economy."[10] But it is without question that Rochet and Tirole were the ones responsible for shifting the vocabulary for what they were describing from the *firm*—which Parker and Van Alstyne still use in the passage just quoted—to *platform*. The significance of this shift is in part that the firm is no longer conceived as a closed box in competition with other such closed boxes, as the company had until then been conceived within management literature and organizational economics. The emphasis on the firm in the first place can also be accounted for by the disciplinary location of these writers, many of whom operate within the subdiscipline of industrial or organizational economics that deals with firms, and the relations of firms to markets.

As Evans and Schmalensee narrate it elsewhere with their coauthor, Andrei Hagiu, "The notion . . . that diverse industries are based on two-sided platforms and are governed by the same basic economic principles is due to a pathbreaking paper by Jean Tirole and Jean-Charles Rochet that began circulating in 2001," and which was officially published in 2003—namely their "Platform Competition in Two-Sided Markets."[11] Rochet and Tirole "showed that businesses such as computer operating systems, dating clubs, exchanges, shopping malls, and video game consoles were two-sided."[12] This article marked an advance in literature on two- or multisided markets, but most important for us here, it also marked the entrance of the term *platform* as a keyword to describe the intermediary object, institution, mechanism, or place that constructs and sustains these markets.[13]

By the time of their oft-quoted 2005 article, "Two-Sided Network Effects: A Theory of Information Product Design," Parker and Van Alstyne had replaced "firm" with "platform," reflecting an important transition in the object of analysis. In this work they note that Rochet and Tirole's 2003 "Platform Competition in Two-Sided Markets," "which focuses on competing credit card markets, parallels our work" and moreover, they continue in an aside, Rochet and Tirole's article cites "an earlier version of this paper, titled 'Information Complements, Substitutes, and Strategic Product Design,' which was presented at the 1999 Workshop on Information Systems & Economics and was posted to ssrn.com in 2000."[14] Yet between their "Information Complements" and their "Two-Sided Network Effects" article, the agent of the economic decisions made about pricing gained a new name: *the platform intermediary.* In their words, in addition to the providers of contents or services and their consumers, there is "a third participant, the focus of our attention here, who produces tools to support both content creators and end consumers. These are platform intermediaries. Examples include Microsoft, Apple, and Sun" (see Figure 1).[15] Despite the reference to tech

Table 1   Two-Sided Market Examples

Product category	Market 1	Intermediary	Market 2
Portable documents	Document reader⊛	Adobe	Document writer
Credit cards	Consumer credit⊛	Issuing bank	Merchant processing
Operating systems	Complementary applications	Microsoft, Apple, Sun	Systems developer toolkits⊛
Plug-ins	Applications software	Microsoft, Adobe	Systems developer toolkits⊛
Ladies' nights	Men's admission	Bars, restaurants	Women's admission⊛
TV format	Color UHF, VHF, HDTV⊛	Sony, Phillips, RCA	Broadcast equipment
Broadcast & publishing	Content⊛	Magazine publishers, TV, radio broadcasters	Advertisements
Computer games	Game engine/player	Ubisoft, ID, valve, electronic arts	Level editors⊛
Auctions	Buyers⊛	E-Bay, Christie's, Sotheby's	Sellers
Academic journals	Articles	*Management Science*	Author submissions⊛
Recruiting	Applicants⊛	Monster.com	Employers
Reservation systems	Travelers⊛	Expedia, Travelocity, Orbitz	Hotels, airlines, rental cars
Shopping malls	Shoppers⊛	Mall of America	Stores
Streaming audio/video	Content⊛	Real audio, Microsoft, Apple	Servers
Paid search	Searchers⊛	Google.com	Marketers
Stock exchange	Equity purchasers⊛	NYSE, NASDAQ	Listed companies
Home real estate	Home buyers⊛	Real estate agents	Home sellers

Notes. This table shows how one side of a two-sided network market receives a discounted, free, or even subsidized good (indicated with ⊛). In general though not always, Market 1 can be interpreted as the user/consumer market and Market 2 can be interpreted as the producer/developer market. We provide a test for which side receives the free good below.

Figure 1. Introduction of the term *intermediary* into analysis. Geoffrey Parker and Marshall W. Van Alstyne, "Two-Sided Network Effects: A Theory of Information Product Design," 1495.

firms here, what is significant is that the intervention of Rochet and Tirole, followed by the work of Parker and Van Alstyne, divorces the term *platform* from its associations with technological infrastructure or base. Instead it is taken to mean a relationship, informed by a particular economic model, and mediated by an intermediary that supports their relationship.

The platform is a channel or mechanism that connects two otherwise discreet and noninteracting sides; it is, as they term it, an intermediary. The significance of this emphasis on intermediaries in the postinternet era— beginning with the commercialization and general consumer availability of the internet in the mid-1990s, which by all accounts was supposed to see the decline of intermediaries and the phenomenon of "disintermediation"— is an issue I will return to below, in the context of Kokuryō Jirō's work. But what is worth emphasizing here is that this third type of platform is *not necessarily* technological. Unlike product-technology platforms and con-tents platforms, platforms as intermediaries are technologically agnostic— shopping malls, bars, credit cards, and magazines are platforms, as much as the technological platforms of game systems and computer operating systems. Moreover, even computer operating systems and game consoles are not "product-technology platforms" in the sense described in the pre-ceding chapter; they are platforms insofar as they coordinate markets—

and hence function irrespective of the code, hardware, or technological conditions that preoccupy research within the rubric of "platform studies" or "software studies," to name two fields that function within the product-technology paradigm.

This "a-technological" sense of platform is highlighted in a widely circulated article Parker and Van Alstyne cowrote with Thomas Eisenmann and published in the *Harvard Business Review* in October 2006, which gives an overview of this strand of platform literature (and was translated and published in the June 2007 issue of the Japanese edition of the *Harvard Business Review*).[16] This article seems to have become the basis of business thinking about platforms (at least until the more recent spate of books on platforms in the mid-2010s, such as *Matchmakers* and *Platform Revolution*). Schmidt and Rosenberg cite it directly in *How Google Works* when they discuss platforms; their conception of the term also implicitly informs Kadokawa Tsuguhiko's account as well, likely due to the Japanese translation of the article. In "Strategies for Two-Sided Markets," Eisenmann, Parker, and Van Alstyne write:

> Products and services that bring together groups of users in two-sided networks are *platforms*. They provide infrastructure and rules that facilitate the two groups' transactions and can take many guises. In some cases, platforms rely on physical products, as with consumers' credit cards and merchants' authorization terminals. In other cases, they are places providing services, like shopping malls or Web sites such as Monster and eBay.[17]

Displacing the firm, the platform is named as the agent that facilitates transactions. Platform becomes *the* principal term to describe the intermediary structure, replacing *intermediary* in the graph (see Figure 2) Eisenmann, Parker, and Van Alstyne include in this article (similar in many ways to the one Parker and Van Alstyne include in their 2005 article). In addition, they specify that platforms are composed of (1) an architecture (including infrastructure or design); and (2) rules, "that is, the protocols, rights, and pricing terms that govern transactions."[18] Key to harnessing the power of platforms is to properly gauge and deploy the "network effects" that drive them—whether same-side or cross-side network effects.

NETWORKED MARKET	SIDE 1	SIDE 2	PLATFORM PROVIDERS
			Rival Providers of Proprietary Platforms
PC operating systems	Consumers	Application developers*	Windows, Macintosh
Online recruitment	Job seekers*	Employers	Monster, CareerBuilder
Miami Yellow Pages	Consumers*	Advertisers	BellSouth, Verizon
Web search	Searchers*	Advertisers	Google, Yahoo
HMOs	Patients*	Doctors	Kaiser, WellPoint
Video games	Players*	Developers	PlayStation, Xbox
Minneapolis shopping malls	Shoppers*	Retailers	Mall of America, Southdale Center
			Rival Providers of Shared Platforms
Linux application servers	Enterprises	Application developers	IBM, Hewlett-Packard, Dell
Wi-Fi equipment	Laptop users	Access points	Linksys, Cisco, Dell
DVD	Consumers	Studios	Sony, Toshiba, Samsung
Phoenix Realtors Association	Home buyers*	Home sellers	100+ real estate brokerage firms
Gasoline-powered engines	Auto owners	Fueling stations	GM, Toyota, Exxon, Shell
Universal Product Code	Product suppliers	Retailers	NCR, Symbol Technologies

*Denotes network's subsidy side

Figure 2. The term *platform* replaces *intermediary* as it was used in Parker and Van Alstyne's prior work, and in Figure 1. Thomas R. Eisenmann, Geoffrey Parker, and Marshall W. Van Alstyne, "Strategies for Two-Sided Markets," 95.

Moreover, despite noting that platforms are not new, Eisenmann, Parker, and Van Alstyne do acknowledge that they have become more prevalent.

Platforms serving two-sided networks are not a new phenomenon. Energy companies and automakers, for example, link drivers of gasoline-powered cars and refueling stations in a well-established network. However, thanks largely to technology, platforms have become more prevalent in recent years. New platforms have been created . . . and traditional businesses have been reconceived as platforms.[19]

As Evans and Schmalensee put it most succinctly, if platforms have existed since time immemorial—marital matchmakers mediating a "market" of potential brides with that of potential grooms are their example of choice—*"The Internet and smartphones have turbocharged* the ancient matchmaker business model."[20] Schmidt and Rosenberg, for their part, also emphasize the technological: "Platforms are increasingly (if not exclusively) technology based."[21] The "not exclusively" is key; I have emphasized that this mode of platform thought is a-technological; however, it is also indebted to changes in communication and commerce that accompany the rise of the internet. If the internet has "supercharged" platforms, it has also put the multisided market on the map as a problem that economists—and some businesses—have until now overlooked, or taken for granted. It has also allowed these writers, as well as businesspeople, to look at their own businesses in a new light.

Platforms after the rise of the commercial internet have become *the* business model of choice. Publisher and media conglomerator Kadokawa Tsuguhiko is a perfect example of this; he emphasizes the importance of rethinking the traditional sites and media of bookstores and magazines as platforms, even as he forged ahead with new places for book selling— the digital platform and book reader BookWalker—and alliances with new media companies such as Dwango, whose Niconico Video site is Japan's response to YouTube.[22] The technologically inspired model of the platform intermediary has indeed allowed businesspeople to reconceive their non-internet businesses along the model of platforms.

I take great care to chart this European and American development of platform theory within industrial economics for three reasons. The first is that it foregrounds some of the most important work around the concept of platforms in recent years, work that media studies had until recently largely ignored (and only now is starting to be cognizant of). It is time for media studies to take this work more seriously, as it offers some clear insights into the operation and economics of the phenomenon of platforms. Second, the prominence of this work within management studies is in stark contrast to a very similar—almost parallel—body of work that developed within Japanese managerial discourse in the field of organizational economics between the years 1993 and 1994, a full decade before the official publication of Rochet and Tirole's foundational article in 2003 (and six years before its informal debut at a conference). Third, an understanding of the work in French and U.S. contexts allows us to draw some contrasts with the Japanese literature on platforms.

Noteworthy in these pioneering efforts on two-sided markets, as well as the Japanese work on platforms from the 1990s, is that this work focuses on platforms as agents of mediation rather than mere technological objects or assemblages. It is for this reason that Negoro Tatsuyuki and Ajiro

Satoshi's 2012 overview article, "An Outlook of Platform Theory Research in Business Studies," dubs this third kind of platform research "interaction-type platform theory," as distinct from the product-technology platforms that occupy much of the research discussed in the previous chapter.[23] Interaction-type platform theory, Negoro and Ajiro write, "focuses on services with functions for intermediation between different users, that is, being a medium for communication or transactions, such as intermediation, settlement, or communication functions."[24] Under this third category would fall internet auction services as well as credit cards and electronic money. The authors have the work of the French and American researchers on multisided markets in mind here, as well as Japanese researchers who emphasized interaction platforms like auctions and credit cards during the 1990s as well. I will rename this *mediation platform theory,* where mediation stands in for all aspects they mention: intermediation, communication, settlement, and transaction.

Network Effects and Technological Conditions

Before moving on to a consideration of platform theory from Japan, we should examine more carefully the relationship between technology and theory in the development of this a-technological mediation theory of platform. If, as we saw Evans and Schmalensee put it, "the Internet and smartphones have turbocharged the ancient matchmaker business model," then it behooves us to consider the *technological conditions* for the development of platform theory in Europe and North America.[25] (Similar conditions motivate the work in Japan, but I will bracket this for the moment.) As it turns out, it is not simply the internet or smartphones that give rise to platform theory but earlier technological developments as well—telephones, the VHS–Beta video format battle, and the battle for market dominance between game console makers. The motivations and conditions for the study of platforms are driven by changes in technology, often *technologies linked to Japanese hardware developments* such as the video cassette recorder (VCR) and third-generation video game consoles in the 1980s. That is to say, the conditions for platform theory are technologically dependent, even if multisided markets are not themselves necessarily technological.

To best grasp this relationship, it would be worth turning to the main source of inspiration for multisided market research. Economic literature from the 1980s on "network externalities" or network effects is one of the main starting points for the work on platforms as intermediaries. Marc Rysman points to the importance of networks literature to that on multi-sided markets, suggesting, "In a technical sense, the literature on two-sided markets could be seen as a subset of the literature on network effects."[26]

As Rysman explains, the two- or multisided market literature is integrally linked to the network literature, a subset of it, even. While distinct in their choice of objects—"media, payments systems, and matching markets" for the platform theory versus telecommunications for the networks literature[27]—the former is dependent on grasping the role of network effects in the adoption of one platform over another (and therefore also in calculating which side of the market to subsidize). Where they differ most is that the objects of network effects literature have high switching costs and require a substantial financial outlay (such as the VHS versus Beta video player battle, where owning both systems would have been very expensive)—whereas internet platforms generally have low to moderate switching costs, enabling "multihoming."[28] Multihoming refers to the tendency to participate in various social media (Facebook and Twitter), or subscribe to several streaming services at once (Netflix and Mubi). Multisided markets are dependent on network effects, which are in turn one of the two key characteristics of the former as Rysman defines it: "A two-sided market is one in which 1) a set of agents interact through an intermediary or platform, and 2) the decisions of each set of agents affects the outcomes of the other set of agents, typically through an externality."[29]

In short, literature on multisided platforms grows out of influential work on network effects concerned with the telecommunications industries and the development, competition, and standardization of hardware technologies. Network externalities "were first defined and discussed in Rohlfs" in a 1974 research paper for Bell Laboratories, note Carl Shapiro and Hal Varian in *Information Rules,* one book among many that repurpose network effects literature for the era of digital economics.[30] (More recently, Richard Schmalensee has emphasized the prescience of Jeffrey Rohlfs's research for considering social media giants like Facebook in an article titled "Jeffrey Rohlfs' 1974 Model of Facebook.") Rohlfs, a researcher at Bell Labs, begins his paper with the simple lines, "The utility that a subscriber derives from a communications service increases as others join the system. This is a classic case of external economies in consumption and has fundamental importance for the analysis of the communications industry."[31] While this work appeared in the context of telecommunications, the greatest developments in this area occurred in the mid-1980s onward, with Michael L. Katz and Carl Shapiro's contributions on network externalities seeking to explain the adoption of a given technology being particularly important. Katz and Shapiro grappled with the problems posed by both computer competing standards (Wintel versus Mac) and the VHS–Beta competition.

The VHS–Beta competition in particular led many to develop and adhere to the premise that "winner takes all" as well as the theory of "first mover advantages" (i.e., the first to market will monopolize it).[32] Let us look, for instance, at the opening lines of Katz and Shapiro's 1986 article, "Technology Adoption in the Presence of Network Externalities": "The benefit that a consumer derives from the use of a good often depends on the number of other consumers purchasing compatible items. Consider, for example, the owner of a videocassette recorder (VCR). At present, there are two competing technologies, beta and VHS."[33] While echoing Rohlfs's introduction a decade earlier, Katz and Shapiro's essay homes in on the problem of the day: Beta versus VHS adoption. Their foundational text on network externalities from 1985, "Network Externalities, Competition, and Compatibility," begins with yet other examples of technologies and standards, opening with computer hardware and software, and subsequently suggesting that the "hardware-software paradigm also applies to video games, video players and recorders, and phonograph equipment."[34] It is significant that one of the starting points and structuring theories for work on platforms as intermediaries is the engagement with the hardware of VHS and Beta, computers, and video game consoles—devices that were by and large made in Japan at this moment in time. As with Japanese automobile manufacture noted in the previous chapter, here too we find the crucial elements of platform theory emerging out of an engagement with Japanese hardware production—in this case, the creation of Sony's Beta and JVC's

(Victor Company of Japan) ultimately dominant VHS video system. This connection suggests a close relationship between hardware production and platform theory.

The concerns of the network externality literature, as well as the influence of the technological systems of the day on its development, continue into the era of platform theory, even as the latter sheds a strict attachment to the technological in favor of a consideration of the mediatory as such. Rochet and Tirole open their pathbreaking article, "Platform Competition in Two-Sided Markets," with the following statement.

> Buyers of video game consoles want games to play on; game developers pick platforms that are or will be popular among gamers. Cardholders value credit or debit cards only to the extent that these are accepted by the merchants they patronize; affiliated merchants benefit from a widespread diffusion of cards among consumers. More generally, many if not most markets with network externalities are characterized by the presence of two distinct sides whose ultimate benefit stems from interacting through a common platform. Platform owners or sponsors in these industries must address the celebrated "chicken-and-egg problem" and be careful to "get both sides on board." Despite much theoretical progress made in the last two decades on the economics of network externalities and widespread strategy discussions of the chicken-and-egg problem, two-sided markets have received scant attention. The purpose of this paper is to start filling this gap.[35]

As we can see from this quote, what Rochet and Tirole call platforms are not the technological objects per se—the video game consoles, credit cards, and so on—but rather the connective tissues and business plans that bind the two sides or markets together. The platforms of the title are physical platforms and formats, as well as more abstract "platforms" like credit card companies. Among the "mini case studies" they examine are credit cards, internet traffic-carrying arrangements, internet portals and broadcast television, as well as four cases from the software industry: the video game market (Nintendo, Sega, Sony, and Microsoft), video streaming software (RealPlayer, Apple, and Microsoft's Windows Media Player),

operating systems (Windows, Apple, and IBM), and text editing programs.[36] Aside from credit cards and some software, most of these case
studies *are* in fact platforms in the technological sense, and as described
in the earlier definitions of the term in the previous chapter. Hence, we
can say that the theory itself is technologically conditioned, dependent to
some degree on systems of mediation; at the same time, it also presents
itself as technologically agnostic, having no particular qualities to the technological system in question, being an intermediary function more than a
technological one. *A-technological* is the best way to describe this disposition: both descriptive of and responding to particular technological systems, and also speaking to the market conditions and business strategies
that allow for the adoption of one system over another.[37] The theory operates with many technological systems even though it concerns a market
formation rather than a technological product. And yet particular technological conditions do favor the creation and theoretical conception of
multisided markets—with digital technologies being particularly prone
to network effects. Not all platforms are digital, but digital phenomena
"produce and are reliant on network effects."[38]

Parker and Van Alstyne radicalized platform theory by taking Rochet
and Tirole's discussion of credit card markets to its logical conclusion; that
is, they posited that all platforms are simply a series of rules or protocols
and the infrastructure to implement these rules, further separating platforms from their technological coordinates. And yet we find the appeal of
platform theory from microeconomics to be its apparent ability to explain
digital economic phenomena. It goes without saying that this was also
its appeal in the Japanese context. The Japanese translation of "Strategies
for Two-Sided Markets" was presented and situated in relation to familiar technologies. Eschewing the English preamble (which opens with the
phrase "Companies in industries such as banking, software, and media
make money by linking markets from different sides of their customer
networks"), the Japanese preamble focuses on technologically supported
mediation platforms, ones known and familiar to readers *as* platforms.
The article opens: "When we look at the business models for PCs, video
games, and cellular phones, we find two different kinds of user groups,
and different value chains and income structures for each. This is called a

'two-sided market' or 'two-sided network,' and these kinds of goods are called 'two-sided platforms.'"[39] The reason for the emphasis on the cell phone as platform will become clearer in the next chapter, but the addition is significant.

The platform *as a-technological intermediary* hence emerges out of a literature very much concerned with how technological ensembles present new market problems—among which are motivations for adoption, incentives, subsidies, and so on. Unlike the literature on product-technology platforms examined in the previous chapter, however, this two-sided market literature comes from a lineage of thinking and research on network externalities, concerning itself with the motivation and effects of adopting technological systems. The problem of mediation is hence more important than technologies per se, a trait that requires us to resituate this work in the lineage of Japanese platform theory, a project to which I now turn.

Japanese Platform Theory

Much of the most vibrant platform theorization in English and French comes from the field of microeconomics or industrial economics, written with the aim of thinking through the problems and challenges posed by multisided markets. Meanwhile, the Japanese theory of platforms gets its start from a more management-oriented group of writers, concerned with the impact of transformations in communications infrastructures on business organization. More interested in the potential disruptions of existing businesses, this work was from early on engaged in the rethinking of the firm and its organization. It is no surprise, then, that the keyword associated with platform in this literature is business—in the transliterated form of "platform business" *(purattofoumu bijinesu)*. As such, this provides in some senses an early account of platform-based business strategy, as well as a proto-account of what today we would call the platform economy—the type of economic and social management that marks the current capitalist configuration (albeit without the critique of these conditions that is so crucial to current work on platform capitalism).[40] Three of the most prominent writers in this vein are Kokuryō Jirō, Kimura Makoto, and Negoro Tatsuyuki, all of whom write about the term coined by Kokuryō—*platform business*—meant to account for internet-mediated forms of commerce, or e-commerce. Negoro and Kokuryō continue to be the leading voices of platform theorization in Japan; they are the two towering figures around which the business-oriented work on platforms develops.

Of the recent spate of books on platforms in Japanese, two of the most significant volumes in the 2010s were published by research teams affiliated with each author. Kokuryō, who teaches at the prestigious Keiō University, edited and coauthored with his Keiō-affiliated Platform Design Lab members the 2011 *Sōhatsu keiei no purattofōmu: Kyōdō no jōhō kiban zukuri* (Platforms for emergent management: Building the information basis for cooperative work). Negoro, for his part, was the general editor of a 2013 volume cowritten by members of his Waseda University research group and the Fujitsū Sōken research and consulting group, appropriately titled *Purattofōmu bijinesu saizensen* (The frontlines of the platform business). Both Kokuryō and Negoro were principal contributors to another significant 2015 volume on platforms edited by no less than the former president of Sony Corporation, Idei Nobuyuki, *Shinka suru purattofōmu* (Evolving platforms), which is part of the Kadokawa Internet Course book series.[41]

Platform Theory in 1990s Japan

This recent spate of work around platforms is in fact preceded by an earlier moment of platform theorization in Japan, in the 1990s.[42] Economist Deguchi Hiroshi offered the first full-blown theorization of the platform as an *industrial formation* in a 1993 article, "Nettowaku no rieki to sangyō kōzō" (Network merits and industrial structure), published in the *Journal of the Japan Society for Management Information*.[43] As the title implies, the article and the argument therein are concerned with the effects of what Deguchi calls "platform industries" *(purattofōmu sangyō)* on existing theories of industrial structure. This article is part of a larger body of work by Deguchi and others that attempts to grasp the potential changes to industrial structure and management practice either as a result of computer technologies and the predicted popularization of the internet (which became available to the Japanese public as of 1994) or inspired by the types of interactions and practices that emerge with the rise of the network society (such as auction sites or the open source software movement).

Deguchi opens his article by noting the increasing prominence of the term *network,* and the fact that for a long time it has been used in relation to public utilities such as electricity, gas, water, and transportation. There, the "hard network" (or physical network) referred to the pipes or roads, and the "service content" to the products such as water or gas being provided to its clients. With information technologies, we have a similar structure, wherein the "network platform" replaces the "hardware network," but in which we also find a separation between platform and the service offered on it.[44] As Deguchi explains, "This concept of platform and the service above it is a layered one."[45] This layer structure is in fact one of the striking characteristics of the computer realm, in Deguchi's view, and moreover it is a stacked one; what is a service in relation to one platform could itself become a platform in relation to a second, higher-level service.[46] Platform provision and platform services are stacked, stratified, or layered.

Another distinguishing characteristic of the platform as distinct from conventional infrastructures is that in traditional networks the platforms and services provided on top of them tend to go hand in hand—the gas companies build the gas lines and provide the gas, the electric companies build the electric lines and run the electricity to your door, whereas within communications and computer industries, the services are often distinct from the platforms they build or run on, and hence a number of services may compete on a shared platform. This phenomenon provides the basis for Deguchi's analytical distinction between "platform provision industries" and "platform service industries."[47] That they can be distinguished does not mean that they are disconnected, however, and Deguchi is quick to point out the effectiveness of examples when platform provision and platform services are integrated or coordinated. The example he gives of the latter is Nintendo's Famicom (or NES) gaming unit with its games, some of which are created by Nintendo but all of which are carefully curated by Nintendo with game producers who publish on the platform going through a rigorous verification process. These are also, moreover, regionally specific and tightly controlled through region-locking the consoles and games. Nintendo benefited both from network effects—stemming from becoming one of the central gaming platforms (here, Deguchi notably refers to Katz and Shapiro's work on the VHS vs. Beta competition)—and economies of scale. As a side note, it is interesting to see the reference to network effects literature here—a body of work that seems to also have impacted the rise of Japanese platform theory, much as it would later impact the rise of French and American platform theory a decade later. This network effects theory was taken up early on the Japanese context as well, with Hayashi Kōichi's 1989 book, *Nettowākingu no keieigaku* (Management studies of networking), being one of the earlier Japanese engagements with this canon, by a writer who was a lifetime employee of NTT, Japan's national telecom.[48]

Deguchi's 1993 intervention is one of the earliest cases of platform theorization in Japan, insofar as it repurposes a technical term used in the 1980s to refer to a type of computer hardware support, transforming it into a conceptual tool used as the analytic basis of "platform provision industries" and their respective (and sometimes vertically integrated) "platform service industries." Yet it is clear from the brief description above that he continues to use the term in a manner both indebted to the layered structure of hardware and software and still very much focused on what can be called hardware and software in the expanded sense these terms take in Japan (wherein the VHS unit or game console is the hard *[hādo]*, and the video tapes or games themselves the soft *[sofuto]*, as explained in chapter 1). In later work, on the contents industries no less, Deguchi would extend his work by reconceiving the platform as a "place" that is a precondition for offering a given service or business—a wider conception that conceives of the service component as a "cultural activity" that requires a "platform," whether this be a stage for a performance or an inexpensive hall for the exchange of fan-made products in the case of Japan's infamous comic markets.[49] Yet even here, in this considerably expanded model of his platform as "place," he still relies on a structurally similar, layered conception of platform and service.

The real break from a technological conception of the platform modeled on the hardware-software paradigm would come in the following year, 1994, when Japanese management professor Imai Kenichi coedited an issue of the journal *InfoCom REVIEW* with his junior research associate Kokuryō

Jirō, a recently graduated PhD from the Harvard Business School, who spearheaded the project. The title of the special issue was "Purattofōmu bijinesu" (Platform business), and the aim of the issue, as Imai writes in his prefatory note, was to provide a conceptual framework that might address the "massive changes the Japanese industrial system is undergoing."[50] The emphasis of Imai's research group, however, was not on computer technologies per se but rather "how the advances and innovations in information and communications technologies lead to changes in the *mechanisms of transactions* between companies, and how these in turn lead to changes in company organization and industrial organization."[51] The focus on these mechanisms of transactions in particular opened up a new front in the study of the platform.

The project of explaining the novel concept of the *platform business* fell to Kokuryō. Acknowledging that most readers probably had never heard of the concept, he defines it thus:

> *A platform business is one where the existence of a foundation or base [kiban, 基盤] provided by a private business allows anyone to supply goods and services to another party under a specific set of conditions, thereby invigorating transactions between third parties and building new businesses.*
>
> Credit cards and other intermediaries of trust allow various businesses to be established, and enable transactions between third parties to take place. Express delivery services, for instance, enable the creation of new transactional forms built around direct-from-the-farm deliveries, allowing the farm owner to establish a profitable business. Or, yet another example of the meaning of platform business can be found in the manner in which Microsoft, by providing what is a "de facto standard" OS, in turn allows for the establishment of independent companies built around offering related products and services.[52]

Each example Kokuryō provides is one in which a basic, standard, or well-defined "base" service or technology provides the ground from which other companies and businesses can spring into existence. What characterizes all of these services—and distinguishes this conception of platform from previous ones—is that they enable transactions between third parties. This is an a-technological theory of the platform as mechanism for mediation in the sense developed above: dependent on certain technological conditions but not technological in and of itself.

Kokuryō's account highlights the facilitation of relations between third parties. This is an aspect that links it more directly to the post–Rochet and Tirole literature that emerges later in France and North America. The emphasis on agents for third-party transactions is at the center of Kokuryō's contribution to the volume, which focuses on what he calls "facilitators of transactions" or "transaction intermediary-type platform businesses" (*torihiki chūkai-kei purattofōmu bijinesu,* 取引仲介型プラットフォーム・ビジネス), a term that he defines in the following manner: "Transaction intermediary-type platform businesses are businesses which invigorate third-party transactions, and which establish themselves as profit-making enterprises by internalizing the added value of these transactions."[53] The particular examples to which Kokuryō refers are Aucnet (a used cars auction network that started in the 1980s over satellite transmission, before moving to the internet in the 1990s), the Visa credit card, and Misumi (an industrial parts supplier and mediator between parts makers and parts buyers). These in turn are the core case studies for the remaining three articles of the "Platform Business" special issue—Hayashi Ryūji's "VISA's International Platform Business," Danno Mikio's "AUCNET's Platform Business," and Nakamura Sadao's "Misumi Corp.'s Platform Business." Given the importance of the credit card to Rochet and Tirole's work, and the Microsoft example as well as auctions to most mediation platform theory, it is striking how the examples Kokuryō and his research associates use are also the ones that would be central to platform theory as it develops in France and the United States in the twenty-first century.

Kokuryō continues to develop his theory of the platform business in his 1995 book, *Open Network Management,* where he looks for further clues into new managerial practices stemming from developments in information technologies. Here he mobilizes the term *open* to signify non-vertically integrated, non-*keiretsu*-based systems, ones that should welcome third-party developers, much as the example of the Microsoft OS.[54] (A *keiretsu* refers to a grouping of firms in a relatively stable, if sometimes informal, relationship, often vertically and horizontally integrated.) He also further elaborates his theory of the transaction intermediary-type platform business, noting that in electronic markets they have to play the following roles:

1. Searching for transaction partners
2. Mediating trust (information)
3. Appraising the economic value of a transaction in an independent [third-party] manner
4. Providing standardized transaction procedures
5. Enlivening trade by unifying various related functions such as the distribution of goods

We might very well say that this is the business of providing the "place" *[ba]* where transactions can take place.[55]

To sum up, we can say the platform business is a "place" (here a metaphorical place or site rather than a physical space) where transactions can occur thanks to the platform's effects of securing a mutual sense of trust between strangers.[56] Indeed, Kokuryō explicitly differentiates this concept of platform from something like physical space by parsing its differences from the term *infrastructure*. While there are some similarities with infrastructure, the emphasis of infrastructure tends to be on hardware, and gives rise to ideas around economies of scale and public good. Platforms, on the other hand, "emphasize the rather 'soft' elements of mediating and adjusting transactions by augmenting a sense of trust."[57]

The importance of trust and the persistence of intermediaries are two of the most significant drivers of Kokuryō's platform research, the academic thrust behind his platform mediation thesis. His time studying in the United States coincided with early discourse around the future of the commercial internet as a site for electronically mediated commerce—e-commerce. Mainstream economists at the time believed that with the rise of e-commerce mediators would disappear, that transaction costs would fall due to the rise of electronic communications, and that industries would experience what was being called "disintermediation." In his classic business text, *The Digital Economy* (published in 1996, contemporaneous with the rise of the commercial internet), digital strategist and evangelist

Don Tapscott glosses disintermediation as follows: "Middleman functions between producers and consumers are being eliminated through digital networks. Middle businesses, functions and people need to move up the food chain to create new value, or they face being disintermediated."[58]

Swimming against the current in this heyday of disintermediation theory, Kokuryō posited that the role of *trusted* intermediaries would in fact be central to any emergent e-commerce initiatives. When the internet replaces face-to-face transactions, he reasoned, the mediating system (the "platform") would be the source of trust, verifiability, and recourse in the case of failed or misleading transactions. In the 1980s, Kokuryō began arguing that in the world of e-commerce, intermediaries would in fact only get stronger, precisely because they would function as the trusted guarantors that the other side would follow through on a promised transaction. This element of trust becomes more important to the degree that platform-mediated transactions replace face-to-face ones. As he recently explained his reasoning at the time: "Because the potential transaction partners would grow in number, the value of trust would become more important."[59] In this sense, his 1989 study of the AUCNET used-car TV-based auction network in Japan was a particularly early example of platform business analysis, *avant la lettre* (that is to say, platform analysis without naming it such; Kokuryō and his coauthor do not use the term in their study).[60]

This prescient sense that trusted intermediaries would become more important over time was borne out in fact, particularly as the commercial internet took root from the mid-1990s onward, and consumer-to-consumer auction services emerged (eBay, Rakuten, and Yahoo! Auctions being some of the major examples in Japan; the success of Amazon's Marketplace is an example of how trust is fundamental to business-to-consumer relationships as well). Kokuryō's work is key to initiating a separation of the platform concept from its prior dependence on hardware or infrastructure. As we will see in the case of Docomo's i-mode mobile internet service in the following chapter, this emphasis on the transactional becomes central to platforms that are more about generating spaces of transaction or mediation than about industrial models of production per se.

Japanese Platform Theory, Stage 2

The next stage in debates around platforms comes in the form of a book published in 1999—the year, coincidentally, that i-mode was launched. Coauthored by two management and business researchers, Negoro Tatsu-yuki and Kimura Makoto, *Netto bijinesu no keiei senryaku* (Management strategies of net businesses) was one of the earliest texts to synthesize and build on existing research on platforms (focusing on the work of Deguchi and Kokuryō in particular). In *Management Strategies of Net Businesses,* Negoro and Kimura point to the importance of Kokuryō's work for the way it shifts to a model of platform as intermediary, or mediation. How-ever, they also wish to push it to be more explicit about the implications for internet businesses. Their criticism of Kokuryō's work is that while it emphasizes the ways mediation platforms become important for internet businesses, it does not sufficiently account for the specificity of internet businesses. Negoro and Kimura focus on what they call, in a riff on Koku-ryō's term, "internet commerce mediatory-type [介在型] platform busi-ness." They define this term in a manner that also builds on and subtly shifts Kokuryō's definition, cited above, writing that it "is a private business that invigorates internet commerce through the mediation of third-party communication."[61] In short, the volume investigates how communication becomes the basis for commerce in the form of internet forums, eBay-style auctions, and online catalog collections. As they illustrate in an important graph, the internet commerce mediatory-type platform business is a sub-set of Kokuryō's larger category of transaction intermediary-type platform businesses (fitting inside this category but separated from noninternet mediated industries such as real estate), and is also a subset of (and hence more specific than) e-commerce in toto.[62] E-commerce includes activities such as the internet-based supply chain management and sales of com-modities through the web that do not necessarily involve third-party inter-actions, and hence would not be considered to be e-commerce mediatory platform businesses.[63]

What is useful about Negoro and Kimura's intervention is its sugges-tion that there is something specific about the form of mediation that hap-pens on the internet, and thus it should be factored into theorizations of the intermediary-type platform businesses. This offers the opportunity to distinguish between a theory of internet as media and the platform as a specific form of mediation. The digital communications-based and subsequently internet-fueled transformations of business were certainly in the background of Kokuryō's theorization of the platform. And he has noted that the specificity of these forms of communication increases the importance of mediators and hence mediatory-function platforms, providing the impetus for his own platform analysis. The internet is powerfully intermediating, bringing people together who otherwise would not communicate, and bringing larger numbers of people into contact than previously. Where Negoro and Kimura differ is in their interest in homing in on internet-specific forms of platforms as mediation, some of which are commercial—Amazon and eBay stand out as both early and precient analyses of commercial platforms that are still in operation—and some of which are noncommercial, such as the Linux and Mozilla open-source software development communities (which they acknolwedge differ from the internet commerce mediatory-type platform business model of the platform insofar as they are not based around the optimization of the supply chain but rather are organized toward the "improvement and development of open-source software through collective intelligence"[64]). As we have already seen, one of the main dividing lines in existing theories of platforms is the degree to which platforms are conceived as layers or sets of hardware, and the degree to which they are seen as mediatory functions—intermediaries in the language of Kokuryō and his collaborators. Negoro and Kimura's focus on internet-based platform businesses notwhithstanding, their contribution broadens the work on platforms as mediatory devices, based on but not determined by particular technologies or media.

To synthesize this account, then, there are several crucial elements of platform theory as we can track it from the Japanese origins of the discourse, through the later U.S. and French developments of the concept: (1) the emphasis on platform as mediation; (2) the emphasis on "place," usually figured as a site of generativity or emergence; (3) the emphasis on social relations as the key to platforms; (4) the emphasis on platforms as a business, hence the formulation of the term *platform industries* (Deguchi) or *platform business* (Kokuryō, Kokuryō and Imai, and later Negoro and

Kimura). These four elements of the platform continue to inform Negoro and Kokuryō's recent work in platform theory.

Genesis of Platform Theory, Revisited

In summary, Kokuryō and his collaborators show a prescient interest in the types of companies and structures that would inspire the subsequent work of Rochet and Tirole in France, and Parker and Van Alstyne in the United States, and which would become the core research around platform business practices—credit cards in particular but transaction sites such as eBay and Amazon as well. Breaking from earlier iterations of the platform as technological substrate—so-called product-technology platforms, usually axed around a hardware-software distinction—Kokuryō, Negoro, Kimura, and Imai use the term *platform* to designate an apparatus of mediation. In so doing they anticipate a similar shift that occurred in the United States and France several years later, in the early twenty-first century. Yet despite the similarities of their work, we may note several distinguishing elements. Whereas Rochet and Tirole and others were concerned with the pricing effects of multisided markets, Kokuryō and his associates were more concerned with the manner in which multisided markets or, more accurately, trustworthy platforms that could be the basis for transactions between third-party individuals or companies, become the basis for new businesses. One group was preoccupied with pricing decisions and market sides, the other with economic effects and institutional repercussions of businesses that often, though not necessarily, operated within industries supported by network communications and internet infrastructure, and built around the transactions between third parties (what in the former literature would become sides of the market). The former situates the platform as a kind of black box that sets the conditions for the transactions that happen on either side; the latter tends to focus more attention on the particularity of the platform involved, including its affordances, the types of activities it allows for, and the types of platforms that exist. To put it a little reductively, the platform research from France and the United States focuses on the edges of the platform, the sides of the markets the platform brings together, while the work from Japan puts more analytical weight on the form of mediation of the platform itself. What binds the two bodies of work together, however, is the theorization of the platform as an a-technological form of mediation, built on a series of standards, protocols, or agreed practices that support novel forms of business creation.[65] The platform is first and foremost an a-technological mediating element that through this *mediation* becomes the basis for new business practices, and new business organization. Hence we might frame this approach as one of *mediation management.* Platforms, in this sense, are apparatuses for the management of relations— economic but also social—allowing platforms to insert themselves into any and all relationships. It is because of this omni-mediatory power that one can arrive at a moment when "platforms are eating our world," as a business book put it.[66]

Regardless of the differences between bodies of platform theory, they have much in common. We must for this reason recalibrate the existing genealogy of platform theory by situating the Japanese theory of mediatory platforms as a key, early moment in the genesis of platform theory. (See Figure 3 for a timeline of publications that sees the Japanese platform theory given its historically prior place in the lineage of mediation platform theory.) Of this there can be no doubt: the genesis of the platform concept, and particularly the mediation platform within executive theory, occurs in Japan. Rochet and Tirole may not have been aware of this work, for which reason we might want to talk about the *plural geneses* or *parallel genesis* of the platform concept rather than a linear, singular genesis. Nevertheless, this new, parallel genesis is significant for several reasons.

On a first level, this new genealogy reorients the historical account of management theory that is of immediate geopolitical consequence. Traditional accounts of theory too often propagate a narrative that flows from West to rest. In this particular case, I take a by-now-hallowed account of contemporary platform theory and research and suggest a need for a thoroughgoing rehistoricization and novel periodization of its development. The flow of theory in this case is not from West to rest but a more complicated emergence from Japan, even if neither Rochet and Tirole nor Parker and Van Alstyne seem to register it (or choose not to engage with it).

At a second level, one might connect this development in Japan to the transnational circulation of management theory; Japan after all may not be part of the "West" as such, but it is part of the Global North and the

Network Effects Literature

Jeffrey Rohlfs, "A Theory of Interdependent Demand for a Communications Service," *Bell Journal of Economics and Management Science* 5, no. 1 (1974): 16–37.	1974
Michael L. Katz and Carl Shapiro, "Network Externalities, Competition, and Compatibility," *American Economic Review* 75, no. 3 (1985): 424–40.	1985
Hayashi Kōichirō, *Nettowākingu no keieigaku* [Management studies of networking] (Tokyo: NTT, 1989).	1989

Platform Theory Literature

Imai Kenichi and Kokuryō Jirō, eds., "Purattofōmu bijinesu" [Platform business], special issue of *InfoCom REVIEW* (Winter 1994).	1994
Kokuryō Jirō, *Ōpun nettowāku keiei: Kigyō senryaku no shinchōryū* [Open network management: New trends in business strategy] (Tokyo: Nihon keizai shinbunsha, 1995).	1995
Negoro Tatsuyuki and Kimura Makoto, *Netto bijinesu no keiei senryaku: Chishiki kōkan to baryūchēn* [Management strategies of net business: Knowledge exchange and value chains] (Tokyo: Nikka giren, 1999).	1999
Jean-Charles Rochet and Jean Tirole, "Platform Competition in Two-Sided Markets," *Journal of the European Economic Association* 1, no. 4 (2003): 990–1029.	2003 (initially circulated 2001)
David S. Evans, "Antitrust Economics of Multi-sided Markets," *Yale Journal of Regulation* 20, no. 2 (2003): 325–81.	2003
Geoffrey Parker and Marshall W. Van Alstyne, "Two-Sided Network Effects: A Theory of Information Product Design," *Management Science* 51, no. 10 (2005): 1494–1504.	2005 (in progress since about 2000)
Thomas R. Eisenmann, Geoffrey Parker, and Marshall W. Van Alstyne, "Strategies for Two-Sided Markets," *Harvard Business Review* 84, no. 10 (2006): 92–101.	2006 (Japanese in 2007)

Figure 3. A timeline of platform theory, integrating the work done by Japanese platform theorists with that done by European and American ones.

tripartite block of North America, Western Europe, and Japan that held for much of the second half of the twentieth century and only began to be complicated around the time period in question—with the collapse of the Soviet bloc, the establishment of the "Asian Tigers," and the increasing industrial and financial might of China. The emergence of Japanese platform theory also comes in the wake of the American and European fascination with Japanese management practice that reached its peak in the 1980s around discourses of Toyotism and just-in-time manufacture (discourses that in turn affected theories of post-Fordism developing at this moment). Indeed, the very term *platform* was inducted into management studies in North America—via Boston in particular—through the MIT-based International Motor Vehicle Program, a research organization that got its start analyzing Japanese automobile production practices, within which the term *platform* circulated early on, in the early 1990s, if not before.[67] Japanese platform theory was itself, as we have seen, impacted by the network effects literature, which was also an impetus for the French and American development of platform theory.

Given this existing engagement with Japanese management theory, we may ask why Japanese platform theory was not immediately taken up. What took so long for Japanese platform theory to be recognized as the progenitor of mediation platform theory as such? Perhaps this returns us to the first point, and suggests persisting imbalances and uneven transfers of knowledge that remain despite Japan's position as nominally part of the Global North. Alternately we can point to the inconsistent and indeterminate work of translation and flows of knowledge—with Japanese car manufacture practice or business organization being more likely to make it overseas than what was then the relatively America-centric internet business models (an America centrism that persists to this day, with all tech hubs inevitably compared to Silicon Valley).[68] Car makers needed Japanese know-how; computer companies did not. Both explanations remind us of the need for a local, regional-specific approach to platform theory and development.

A third and final reason for the significance of this Japanese platform genesis is that the development of mediation platform theory had immediate and appreciable effects on the development of one of the most important cultural-technical-business events of our time: the invention of the proto-smartphone—the "feature phone"—in the form of Docomo's i-mode system in 1999. This development would put Japan on the map for the telecommunications industry, but despite the many excursions of journalists and tech company heads to Japan to figure out how i-mode worked, the platform *theory* that anticipated the mobile device was not brought back to the United States except in the form of platform practice. The i-mode system is both an immensely successful implementation of the transactional theory of platforms and a forerunner of everything to come in the smartphone present. This is to say that the genealogical origins of the iPhone do not lie in Steve Jobs's brilliant mind or Jony Ive's design sense, as the story is so often told, but in the economic relationships, transactional setup, and technological affordances offered by the platform theory outlined above, and put into practice by the iPhone's precursor in name and game: i-mode, to which I turn in chapter 4.

Coda: Toward the Platformization of the Economy

Before concluding this chapter, however, I would like to make a final point about the wider impact of this strain of management theory, and the ways this strain of mediation management theory expands the purview of management as such. Traditionally, management had been concerned with the labor force under its control. As Harry Braverman details in his classic text *Labor and Monopoly Capital,* management formatted the laborers and their activities, in the service of the corporation they were employed to work for. With the emergence of mediation platform theory we see a blurring of the lines between industrial organization and social organization in its totality (what we might call "society" or the "social field"). It is at this moment when we see foreshadowed the platformization of the economy, or the beginnings of a transition from economy to the platform economy. Of course, *readings* of Fordism such as Antonio Gramsci's emphasized the impact of the conception on the totality of social life, including life outside the factory. My claim here is that this impact on all social life is intrinsic to mediation platform theory.[69] This is perhaps the most profound shift that platform theory brings to the field of corporate and economic management *and* to society on the whole. Management is no longer content to develop efficiencies in the organization of the firm; the efficiencies of the firm are now fully dependent on the corresponding organization of the social body in relation to the firm. By situating itself as a mediating operation between company and other organizations via the conception of platform as market creator, platform as management theory reorganizes the entire social field. By being the mediator or the "middle," the platform operator asserts power over all entities it mediates—whether these are other companies, or users, or by means of the users, society itself. (As we know too well, the term used by Silicon Valley to denote this transformation is *disruption*.[70])

Platform theory hence results in a total mobilization of all companies or market sides around the platform as object that mediates, and the company creating the platform, leading to a transformation of the consumer (or user) as much as a transformation in a given organization. Digital media theorists have rightly figured users as producers within platform capitalism. This is not simply because they produce content or data for the platforms they interact with but also because in so doing they literally form part of the production process, *at the very moment they enter into one of the multisided markets that platforms generate.*[71] Users become coproducers of the other parts of the market, coordinated and managed by the owner of the platform. Management thereby expands its purview and its hold, mobilizing the entirety of the economy and society as part of its mandate.

The managerial strain of platform theory's most lasting effect is this expansion of management's control from the workers to products (here being *contents*) to the consumers themselves, borne out in "end-user license agreements" (EULAs), terms of service (ToS), and other forms of digital rights management (DRM) agreements that any user of a computer or mobile device or service must now confront, and that ensures that contents forever remain packaged intellectual property (IP).[72] Through the justification that they form part of the market under control, the operation of management now extends to users themselves. Mediation platform theory is the Trojan horse by means of which the total management of all sides of a market by one side of the market is theorized and justified. The network effects that lead to one platform corporation's monopoly leads to the monopoly of power and control over all sides of the market—something we see so clearly with Facebook, Google, and Netflix, and find prefigured in the management of all parties by Docomo's i-mode before them. It is this expansion of managerial power through the principle of market mediation that most justifies the term *platform capitalism,* which draws attention to the pervasive power of platform's mediation as it operates through the three moments of production, circulation, and consumption, transforming economy and society in its wake.

CHAPTER FOUR

Docomo's i-mode

Formatting the Mobile Internet

Over the last three chapters, we have seen the conceptual genesis of the platform concept in its various forms, as well as the history of its sister term, *contents*. But this conceptual and discursive genesis must be supplemented by another, parallel narrative, which revolves around the development of the platform *in practice*. If on the one hand we have the discursive genesis, on the other we find the practical genesis of the platform concept, which also features the integration of contents and platform together for the first time. To locate this development and implementation of the platform, we turn to i-mode, where the platform concept was put into practice in a manner that would have repercussions for the smartphone to come, and the commercial internet as a transactional space more generally. I-mode as platform names simultaneously a technical object and a market construct, unifying the three types of platforms described in the preceding chapters into a single, monetizing platform paradigm, which in turn created a contents market.

I-mode is the name given to an internet service for mobile phones, launched by Japan's largest mobile provider and former telecommunications monopoly, NTT Docomo, in February 1999. Its immediate and immense commercial success made it and Japan the world's foremost example of a successful mobile internet ecosystem. At the turn of the millennium onward, writers around the globe effused about the mobile ecosystem of i-mode and its two competitors: au by KDDI's EZweb and J-Phone's (later

Vodaphone and then SoftBank) SkyWalker and SkyWeb services. Here I focus on i-mode, as it was the pioneer and the model for these other services; it also, for a long time, dominated market share, having the highest number of users in Japan. As we will see here, i-mode was a platform in all three senses detailed: a technological substrate, a mediator between contents layer and the platform layer, and a facilitator of transactions between third parties. There were platforms in Japan and elsewhere before i-mode, but this was perhaps the first time such a technological market-mediating service was created as and called a platform—or, in the words of one of the i-mode creators, Natsuno Takeshi, as a "platform concept."[1]

If this starts as the story of i-mode, it is also of necessity a narrative of the "i-mode effect"—that is to say, of the profound impact i-mode as platform had on the contents ecosystem in Japan, on the formation and development of the global smartphone (represented today by the two dominant operating systems, iOS and Android), and ultimately on the development of the mobile and fixed internet landscape in Asia. The analysis of the i-mode effect will spill over into subsequent chapters and should be considered a key component of any narrative of the smartphone—even if it is too often neglected in "smartphone story" volumes, which frequently divorce the iPhone from the longer history of smart mobile phones and their ecosystems on which the iPhone itself was modeled.[2] As Elizabeth Woyke writes, "Japan also sparked the world's first mobile media revolution, in 1999, via DOCOMO's mobile Web service, i-mode."[3] For her part, mobile media scholar Larissa Hjorth early on and in relation to i-mode in particular positioned the cell phone as "a multimedia device par excellence—a plethora of various applications that operate across aural, textual and visual economies"[4]—that interfaces directly with discourses and practices of media convergence. Today, she would argue that i-mode was a smartphone as we know it in all but name.[5] Gerald Goggin notes that the iPhone "shares more with previous cell phones than Apple concedes," suggesting that "the iPhone is actually a fascinating case of what can be called 'cultural adaptation.'"[6] One of the main sources of inspiration and objects of adaptation for the smartphone was, this chapter will suggest, i-mode. Natsuno Takeshi, a key architect of i-mode, writes that with i-mode, the mobile phone became a true multimedia device.[7] This chapter will sketch the outlines of the i-mode project, the development and articulation of this multimedia system, and the transactional relation between the i-mode platform and contents that is embedded within it. In doing so, I will discuss its instigators (McKinsey & Company consultants among others), its closest platform precursors (AOL and Nintendo in particular), and the manners in which it models the multiple articulations of platform we have seen over the previous chapters.

I-mode was a system like America Online (AOL) was for the wired internet: a walled-garden model of the web experience, this time catered to mobile phones. Mobile devices enabled with i-mode were feature-rich phones (hence their name, "feature phones") that brought users the ability to access specially formatted web pages (a variant of HTML known as Compact HTML or cHTML). Pressing the centrally located "i" button (generally located in the center of the menu buttons on clamshell [flip phone] model cell phones) launched the browser and took one to the main menu, from which one could access a set of services from banking and plane reservations (which were free) to contents services, for which one paid—such as weather, news, ringtones, wallpapers, games, and so on. While now a distant memory, ringtones were the first major market for paid contents services, and a significant source of income for the music industry in the early days of cell-phone-mediated commerce.[8] Subscription-based games are another example of popular contents. These services were presented along the model of early web portals such as AOL, with the four main categories of services and subservices arrayed beneath them. Hyperlinks took one from the main portal page to more specific pages, until one finally arrived at the contents one was searching for. Official i-mode web pages were subjected to rigorous vetting process by Docomo, and any contents that could be paid for via Docomo's central fee-collecting services required official i-mode status. Unofficial i-mode pages existed—especially for pornographic contents but also pages made by hobbyists and others—but these would not be accessed via the central menus. I will return below to the specifics of the i-mode system.

For a number of years, i-mode was regarded as the future of mobility. Article after article appeared in English-language trade press and popular magazines—particularly in the early twenty-first century—with titles such as "I-mode Is Coming," "DoCoMo Rising," "Coming to America," "Lessons from Japan: U.S. Operators Can Take a Page from the NTT DoCoMo Book," "On Top of the World" (in which *Fortune* named i-mode creator Matsunaga Mari number 1 woman of the year in Asia for the year 2000), "Japan's Wireless Wonder," and "Internet a la i-mode."[9] I-mode was the mobile future other countries were urged to emulate and follow.

Not surprisingly, i-mode has been subject to several important analyses and many brief mentions within platform literature—most notably within the product-platform context, including Gawer and Cusumano's chapter on the failed efforts to internationalize i-mode in their *Platform Leadership*, and as an intermediary within David S. Evans, Andrei Hagiu, and Richard Schmalensee's *Invisible Engines*.[10] However, these have situated the service in relation to their respective business models, and in the former, solely in relation to its efforts to internationalize. Important communication studies work like Ken Coates and Carin Holroyd's *Japan and the Internet Revolution* situates i-mode in the context of communications technology in Japan; and Mizuko Ito, Daisuke Okabe, and Misa Matsuda's edited collection *Personal Portable Pedestrian* examines the immense cultural impact of i-mode, focusing in particular on the lives of its users. Here the aim is to offer an overview of i-mode in relation to the platform literature discussed in earlier chapters, showing how it operated as a practical development of the platform concept. One of the things we will be paying attention to is the symbiotic relation that develops between contents and platform, ingraining an understanding of the mobile market as the foremost site of platform innovation. The eventual eclipse of i-mode by the iPhone and Android smartphones in the early 2010s is in this sense proof of the i-mode platform concept as much as a signal of its demise; the long-term effect of the system on the Japanese, Asian, and indeed global configurations of the mobile ecosystem cannot be underestimated.[11] I-mode played a decisive role in formatting the global experience of mobile media; it affirmed the closed model of the internet as a series of pipes to walled-garden sites or apps as the default of mobile operating systems. Google CEO Eric Schmitt openly told Natsuno that he wanted to take the i-mode concept and extend it to the world—and given Android's global dominance (commanding roughly

88 percent of the global market share for smartphones), it is safe to say his company succeeded in this ambition.[12]

Docomo's i-mode set the cultural conditions for e-commerce and continued to affect the very model that start-ups and new ventures in Japan and elsewhere in Asia took over the decade and a half that followed. By the same token, i-mode was a platform for the development of further platforms, a seed bed for later platforms, and a site where contents providers flourished and later became platforms in their own right.[13] As such, i-mode led to the emergence of the companies that would lead the way in the second phase of platform production, from around 2004–5 onward, which survived i-mode's demise, and which I will analyze in greater detail in chapter 5 and the conclusion.

Mobile Internet Conversion

Launched in 1999 at a time when internet penetration was very low in Japan, i-mode was many people's first encounter with the web. As such, it also formatted the experience of the internet as one based on paid services—quite different from the PC model of the web that was dominated by the ad-funded and venture capital–fueled "free services" ethos, particularly dominant in Silicon Valley, that ultimately led to the transformation of data itself into commodity as these firms sought a source of income.[14] The rollout of i-mode was fast and had an almost instantaneous adoption, fueled by the marketing techniques of Docomo, the high renewal rates of cell contracts and phone purchases, the salespeople in the Docomo-run stores, the low cost of i-mode-enabled cell phones relative to non-i-mode phones, and the low monthly cost of the service itself.

By the end of 2002, a full 80 percent of cell phone users subscribed to mobile internet services (a figure that includes i-mode's competitors, KDDI and J-Phone); by comparison in the United States, only 8.9 percent of cell phone users had mobile internet in 2002.[15] Docomo held the lion's share of these users, at a 64 percent market share of the mobile internet market.[16] Moreover, by September 2000—a year and a half after its February 1999 launch date—a full 90 percent of Docomo wireless subscribers were also subscribed to i-mode.[17] No wonder the reams of articles on Docomo at the time saw i-mode as the future of the mobile internet. Indeed, this was a future that Docomo attempted to realize with significant investments in U.S. and European telecoms in the early twenty-first century. But despite the high hopes and the fervor of journalists, who wrote as if peeking into Japan was effectively peeking into the future of the global market, these ventures were ultimately not quite successful.[18]

Why did so many Japanese consumers convert to or adopt mobile internet so quickly? Docomo's finesse in making the internet accessible and palatable to large numbers of people from a diverse range of age groups was one of the central elements of its success. In part, this success was based on the rather expensive—in some ways prohibitive—costs of internet access in Japan. While there were internet providers, NTT charged fixed rates for the mere use of the telephone line during the daytime. Not only were users of dial-up required to pay the internet provider; they also faced additional charges on time spent online, calculated by the minute and paid to their telephone providers. By one calculation, "Japanese dial-up Internet users paid three times U.S. rates for an hour of access."[19]

I-mode's packet-based system allowed users to get online easily by way of their mobile phones—no technical expertise required—and only pay for the number of packets switched rather than the time spent online. Moreover, the low monthly fee of 300 yen (US$3) charged for adding the i-mode service to one's phone led many to adopt the new service based on the low cost considerations—even if they did not plan on using i-mode or had only a vague idea about what i-mode was. From its launch in February 1999, to 2001, i-mode's number of users soared to more than 36 million, a figure that was higher than AOL's *worldwide* subscriber base, which had just topped 33 million at the end of 2001.[20] In short, in only two years since launch, i-mode's success had turned Docomo into the largest Internet Service Provider (ISP) in the world—even if its entire subscriber base was located in Japan.[21]

Another reason for this quick adoption of i-mode was the rather rapid hardware turnover rate already established in the late 1990s for mobile phones in Japan. Customers were used to switching their phones regularly. When it was time to switch to a new phone, i-mode-capable phones were promoted by Docomo staff in their stores. Point of purchase hard sell was a significant factor in the widespread adoption of the service, and the i-mode crew took store staff education and proselytizing seriously in their planning for the service rollout. Docomo's quasi monopoly position in the mobile market (accounting for 60 percent of the mobile market share) was yet another reason for the success of the i-mode rollout. Indeed, we can think of the hardware turnover rate and the low access fee as passive reasons for i-mode's success, driven not by need but by the classic tactic of persuasion. For this to be effective, Docomo made sure i-mode-equipped phones were at the same price point as non-i-mode phones, and the i-mode team had priced the monthly access fee so low that anyone would willingly tack it onto their monthly bill. This perceived low cost of service also extended into the subscription to contents as well, also priced along a 100- to 300-yen range (US$1–3). The price of $3 per month for access to what had become one's favorite game seemed cheap in relation to the much higher sticker cost of console games, even though the yearly cost of $36 for a basic mobile game may make one wonder whether it was indeed as cheap as it seemed (this subscription cost vs. sticker cost ratio has echoes in the freemium game model today). As a sidenote, Docomo deployed a number of "tricks" that prevented people from easily canceling a service once subscribed; one could not unsubscribe at certain hours of the night (when Docomo engineers found the largest number of people tended to unsubscribe), and anecdotally one hears about the numerous menus and pages one had to dig through in order to actually reach the unsubscribe option.[22]

Yet simply being signed onto i-mode did not make it a success; using it did, and there were enough avid users to make it successful. To understand why, we have to look more closely at the conception and development of i-mode's platform concept and contents strategies.

The i-mode (Eco) System and the Platform Concept

I-mode was a revolutionary system that introduced Japanese cell phone users to a new kind of phone: the internet-enabled proto-smartphone that could send and receive email, reserve plane tickets, and make bank transactions. While definitions abound, the smartphone is in essence a pocket-sized computer that is also a pocket supermarket for media commodities,

the movement of which is restricted by digital rights management en-
coding that prevents the user from transferring particular contents from
the device elsewhere. The epitome of portability, the device also imple-
ments and manages importability—particularly in its most restrictive form,
Apple's iPhone. And yet the importability we find in the iPhone was nothing compared to the lock-in of the feature phone. Woyke writes: "A number
of people think [the smartphone] was born in 2007, when Apple cofounder
Steve Jobs proudly showed off the first iPhone at the Macworld conference
in San Francisco. But what many people either forget or do not know is
that phones with smartphone features had already been on sale for more
than a decade."[23] Indeed, insofar as i-mode fits the definition of the smartphone as a "cell phone that includes several software functions, such as
e-mail and browser,"[24] it is arguably one of the earliest *mass scale* rollouts
of the smartphone as such, suggesting the only substantive difference between the "feature phone" (as prior phones, including the internet-enabled
clamshell phones in Japan, were called) and the smartphone were form factor and touch-screen interface (flat rectangular slab of touch-screen glass).

The three architects of i-mode were Enoki Kei'ichi, Matsunaga Mari,
and Natsuno Takeshi.[25] Enoki was a lifelong NTT employee tapped to head
the i-mode project; he hired Matsunaga, the founding editor and editor in
chief for a number of magazines run by the massive job-recruiting firm
Recruit. Matsunaga in turn brought internet entrepreneur Natsuno into
the i-mode project. At the time Natsuno was an executive at an internet
start-up offering free dial-up internet access in return for an ad-mediated
service. Natsuno had experience in the tech industry, and Matsunaga had
experience producing contents. Enoki was convinced that the key to the
i-mode system would be its contents; this is what would attract users and
keep them hooked—hence his decision to recruit Matsunaga, a self-declared
technophobe, who would become something like the editor in chief for
i-mode contents. There she developed a vision of i-mode along the model
of the concierge—existing to provide select information for its users.

As Natsuno Takeshi writes in *I-mode Strategy,* "I-mode is a business
model, not the name of a system."[26] We might paraphrase Natsuno's statement that i-mode was not a technology; it was a business system, a "mechanism," as Natsuno writes of it elsewhere.[27] Natsuno's main job was to recruit

contents providers—companies that would develop pages, services, and contents for i-mode. These services could either be free—such as banking and train reservation services—or paid. Contents had to be paid for; Docomo did not recognize the "freemium" model of "free to play but pay for 'premium' goods" so popular in games today.[28] Again, it is worth emphasizing how different this pay-to-play model was from the PC internet, whose users expected everything to be free (and for which we now pay with our data).[29] Services that charged generally billed on a subscription basis, renewed automatically each month. All fees were paid through Docomo, as part of one's wireless bill. Docomo, in turn, charged contents providers 9 percent on every transaction for the trouble of collecting their fees (one might compare this to the 30 percent cut currently taken by Google Play and Apple's App Store). The business model of taking a 9 percent cut came from an earlier Docomo telephone information service, "DialQ2," that it started in 1989, according to former Docomo employee and i-mode project participant Kurita Shigetaka, best known as the inventor of emoji.[30] That said, the main income source for Docomo's i-mode came from network access fees (300 yen / US$3 per month), and most significant of all, data fees, on which it made a killing. Indeed, Evans, Hagiu, and Schmalensee note that "NTT DoCoMo's i-mode mobile Internet platform also earns a disproportionate share of its profits from end users, in this case through network usage charges. We have seen that DoCoMo also earns some revenues from 'official' contents providers who choose to use DoCoMo's billing system, but this accounted for only around 1 percent of total revenue from users in 2004."[31]

Contents companies prospered too, receiving a steady income and encouraging other contents providers to join the fray. A good number of these went public on Japan's recently created tech stock markets, Mothers and JASDAC. (An equally large number later went bust as i-mode went into decline in the 2010s, after the introduction of the smartphone and the shift to Apple and Google as the main mediators of contents via their app stores.) A virtuous circle was created between handset manufacturers, which continually introduced new features; i-mode users; and i-mode contents providers. In Natsuno's telling, this emphasis on contents "created a so-called 'positive feedback' cycle, the same that drives the Internet itself—

namely, good content attracts users, which in turn attracts more content, and so on."[32]

This feedback cycle in turn requires a modification in the value chain model used in business studies. There, the Michael Porter model of the value chain had reigned supreme. According to Porter's diagram, the creation of value within a firm moves from left to right, with the farthest right being the one-time intersection of the firm with the consumer via the market (see Figure 4). In other words, this model posits one "side" or market, the moment the firm sends its completed product out into the world and into the hands of the consumer.

By contrast, writes Natsuno, i-mode requires a different model of the value chain, one in which there is a circular relation or series of feedback loops between multiple participants in the i-mode project.

Mobile phones with sophisticated features but that are easy to operate; a telecommunications network to link those phones; portals that enable

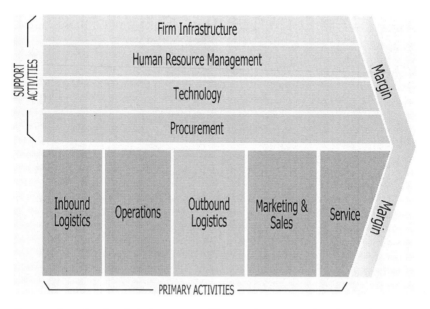

Figure 4. Porter's value chain. Image created by Dinesh Pratap Singh and posted on Wikipedia.

subscribers to find the content they want easily; and a business model that supports the distribution of that content: a value chain that delivers greater user satisfaction is formed through the interactions of these platform and content elements.[33]

These elements of the platform interact in a multidirectional manner; the value within this value chain emerges from all the interactions, moving in multiple directions simultaneously, as Natsuno shows in a graph accompanying his description of i-mode in *The i-mode Wireless Ecosystem* (see Figure 5). This is an image of a value chain based on multiple markets interacting simultaneously—in the language of the previous chapter, a multisided market.

Natsuno describes i-mode best when he writes, "Mobile and Internet technologies alone did not make it a success. Content providers, manufacturers, Docomo, and our subscribers all interact, influencing each other to create this new service; they are all part of the value chain."[34] Implied here is a model of the mediation platform, as he makes explicit in the following passage.

We leave the content creation to the service providers who excel at that; Docomo concentrates on our system for collecting fees, our platform, and

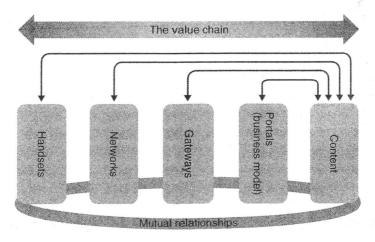

Figure 5. The i-mode value chain. Natsuno Takeshi, *The i-mode Wireless Ecosystem*, 4.

designing our data warehouse. This is what we might call our platform
concept [*purattofomu no hassō,* プラットフォームの発想]. Docomo's role,
on which we focus, is to provide a win-win platform.[35]

Docomo, in other words, focuses on providing the mediatory platform
between users and contents and handset makers, encouraging mutual dependence by all parties and creating a virtuous circle of dependency and
consumption among them.

In his first book, *I-mode Strategy* (published in Japanese in 2000 as
I-mōdo sutorateji), Natsuno uses chaos theory to describe the uncertain yet
generative relationships between the different players in the value chain;
in his second book he adopts the metaphor of the ecosystem. The "ecosystem" has since become a ubiquitous term in mobile circles, presciently
(and possibly influentially) deployed by Natsuno in a 2002 book translated
into English in 2003 as *The i-mode Wireless Ecosystem.* Natsuno's usage
was itself likely influenced by the work of business strategist James E.
Moore, who first suggested the concept of ecosystem as a way of grasping
new business models, particularly those related to the tech sector, in a 1993
article and later in a 1996 book titled *The Death of Competition: Leadership
and Strategy in the Age of Business Ecosystems.*[36] However, Natsuno's use of
the term in relation to the mobile ecosystem is the first I have found in that
context, possibly securing this concept's pride of place in current discussions of the now familiar refrain around the mobile media ecosystem.[37]
Indeed, in a later book, Natsuno explains his choice of the ecosystem metaphor: "The success of i-mode lay in the creation of an active 'ecosystem'"—
which included not just Docomo but also "contents business operators,
mobile handset manufacturers, server manufacturers, and users each bringing together their knowledge."[38] In short, the ecosystem concept is integrally
linked to the conceptualization of a multisided market or platform that
brings together distinct parties. It is presumably the ecosystem concept's
natural affinity to a holistic view that allows this term to come to prominence within roughly the same time frame as the platform concept. Yet it
was the Japanese mobile carriers who developed this ecosystem concept—
European and American carriers did not share Docomo's ecosystem vision,

which is why the mobile internet had not spread to those countries, according to Natsuno.[39]

The graph in Figure 5 that shows the interlocking elements of the i-mode system and indeed the very concept of ecosystem demonstrates Natsuno and his colleagues' understanding of i-mode as a platform *as* intermediary, a mode of mediation between multiple stakeholders that is the very definition of the platform as theorized by Kokuryō, Negoro, and Kimura, and later Rochet and Tirole and Parker and Van Alstyne. This is also a point later made by economists David S. Evans, Andrei Hagiu, and Richard Schmalensee in their 2006 book, *Invisible Engines: How Software Platforms Drive Innovation and Transform Industries,* where—drawing on Natsuno's description of the i-mode phenomenon—they also note that i-mode has all the characteristics of a multisided market; it acts as a platform where consumers and producers of content meet and transact, but it also has a third side, the handset makers, making this a three-sided market.[40]

All of which brings us to the question, how did the i-mode platform concept emerge, and was it informed by the work of Kokuryō Jiro? When I asked Natsuno about the platform concept of i-mode and whether he was drawing on Kokuryō's account of the platform from the mid-1990s, Natsuno claimed he was not; he was in the United States at the time, so there is no way he could know of it, he says. His concept of the platform came from his experience in urban planning, Natsuno notes: the more you bring services together, the more you can bring people together.[41] Kokuryō was in turn hired by Natsuno at one point in the early twenty-first century, at a time when the government was thinking of taking antitrust action against Docomo to break up their perceived monopoly in the mobile internet market. Kokuryō cannot say whether Natsuno ever read his account of platforms—and yet he too approached platforms as a means of differentiating them from infrastructure, and the forms of monopoly or exclusivity that inhere in infrastructures.[42] Perhaps, at that point in time, arriving at a conception of the mediation platforms in Japan was a natural evolution from an engagement with infrastructure and an extension of thinking about the impact of digital technologies on business practice for both Kokuryō and Natsuno. Regardless, the similarities in their conception of platforms is notable. In a book that explains his work smarts, rather boldly titled *The Work Techniques of the Man Who Made One Billion Yen,* Natsuno writes:

> It was Matsuda Mari who first gave me the nickname "infrastructure geek" *[infura otaku].* It's true, I'm an infrastructure geek. I'm the type who finds platforms—in other words, making things like foundations [*dodai,* 土台] or bases [*kiban,* 基盤]—fun. Platforms, then, are both a foundation for something, as well as the above sense of mediation.[43]

Echoes of Kokuryō's work are palpable here, whether conscious or not, as we can see the two meanings of the term as mediation between multiple markets and the foundation for other economic and social activities found in Kokuryō's work to be at the core of the i-mode project and its platform concept.

That said, Natsuno, Enoki, and Matsunaga were not the only platform strategists involved in the creation of i-mode. Another set of key players in this story were consultants from McKinsey & Company, who functioned as agents of the globalization of mobile technology, strategy, and business practice.

McKinsey & Company

By the mid-1990s, the idea of a cell phone linked to the internet was already being widely floated and was assumed to be part of the next wave of wireless communications. But the company that initially proposed the i-mode service to Docomo and was most closely involved in its concrete realization was McKinsey & Company, a consulting firm with a global reach that had established offices in Japan in 1971—at a moment when the Japanese corporate model seemed to be eclipsing the American one.[44] While Docomo had already developed the "DoPa" system of wireless data communications, it had not gained a market. McKinsey was there to propose ways to bring the technology to market successfully.[45] The company proposed "the concept of a 'new mobile phone for beginners.'"[46] At first the idea was to use existing Short Message Service (SMS) texting technology to distribute key information for a price; Matsunaga was brought on board with the plan to have her curate this content, given her expertise with magazine editing. With the saturation of the SMS market, Enoki and the i-mode project pivoted to using the mobile internet as a distribution mechanism. The technology was already mostly in place; what Docomo needed was a rollout that would make this technology palatable and attractive to users. Content was crucial in this regard, and Enoki "believed that content was the key to the new project."[47] The major question became, How would Docomo provide this content? And who would decide what would be provided?

As Matsunaga and other i-mode creators note, McKinsey & Company's Tokyo unit proposed the very concept of i-mode to begin with—or at least its outlines. The specifics had to be filled in by the i-mode staff, as project leader Enoki emphasizes. In Enoki's retelling of it, a forty-page document was handed to Docomo top brass by McKinsey. The essence of the report was that "Docomo should use its existing customer base—in other words, the number and financial potential of its customers—to quicken the circulation of information by positioning itself between its customers and contents."[48] The McKinsey report essentially suggested that Docomo create a platform for the distribution of contents to consumers. The report added that Docomo "would gain financially by the growth in data traffic, as well as gaining money from the growth of non-traffic-related income streams, namely the sale of paid information services as well as advertising fees."[49] For McKinsey, the aim was to monetize the provision and flow of information. However, Enoki emphasizes, while the report outlined the general objectives, it had no details to offer about the specifics of the product or how it would be sold. Moreover, while the report offered business examples of the fixed-line internet, it had no specific advice for how to develop a mobile internet service or how the latter might differ from the former.[50] Still, the report was, by all accounts, the impetus for the project that resulted in i-mode. Docomo's top brass understood well that the market for cell phones that merely made calls would quickly be saturated, and that data was the obvious next phase for the Japanese mobile industry. It was the McKinsey report that suggested the contours of this future.

NTT Docomo president Oboshi tasked Enoki with creating the type of data service outlined in the McKinsey report when he promoted him to the then newly established Keitai Gateway Division. Enoki was the first employee in the division, and his first order of business was the hiring of other staff. There were two other McKinsey staff, however, who participated in the i-mode development and rollout. In a departure from McKinsey's usual mode of doing things—analyzing a company, creating guidelines for its restructuring, and leaving the rest to the company to execute—it did not just leave Docomo with the report as soon as it had been completed.[51] Rather, the staff stayed on at Docomo's request, to oversee the implementation. The Docomo implementation team for the i-mode project hence included embedded McKinsey employees. This was unusual for McKinsey, but it was unusual for the Docomo employees, too, as is retold in the multiple accounts of conflict and generally strained relations between the McKinsey crew and the central players on the Docomo side. For instance, the McKinsey side apparently disagreed with the hiring of Natsuno—who became one of the most important figures in the i-mode success story— and also had conflicts over whether Docomo should produce its own contents or not. The McKinsey side argued that it should, believing that "product content is high risk but has high returns." The premise was that a single hit contents would lead to significant financial benefits for Docomo.[52] The Docomo side—Natsuno and Matsunaga in particular—disagreed, and thought that other companies would not step in and produce their own contents, as was the plan, if they knew they would be competing with Docomo in the realm of contents production. Natsuno and Matsunaga argued that these companies would presume Docomo would get home team advantage and promote its own contents above all others. These and other issues became a source of tension between the McKinsey analysts and the Docomo employees.

Regardless of how well they did or did not get along, the McKinsey consultants were present for much of the development of i-mode, and they can be assumed to have had a significant role in its development. They also played roles of varying importance in Japanese internet culture in the years ahead. In Matsunaga's account of events, she notes the McKinsey consultants' tendency to sprinkle English words throughout their strategy discussions—"technical jargon and McKinsey speak," she calls it.[53] "They spoke more English than Japanese," she continues, "and were extremely difficult to follow." While Matsunaga singles out MBA terminology such as *logic tree* and *strategy,* she also points to their use of terms like *on demand* and *users*—terms that would become the mainstay of Japanese internet discourse.[54] But this importation of English terms raises questions about the status of importation or translation in the creation of i-mode. So let us look for a moment at who these consultants were.

Three of the main McKinsey people involved were Nanba Tomoko, a McKinsey partner who officially led the McKinsey i-mode team during the project but was not closely involved in daily activities; Yokohama Shin'ichi, who became a resident McKinsey personnel early on in the project and worked for two decades at McKinsey before moving to NTT Data; and Obara Kazuhiro, who was also a more junior but resident McKinsey personnel for part of the project. Nanba left McKinsey in 1999 to found a mobile start-up, DeNA (pronounced "D-N-A") which focused at first on mobile-based auction sites for i-mode's mobile internet competitors, au (KDDI) and J-Phone (Vodaphone/SoftBank)—which together occupied something around 40 percent of the market share and did not meet with the immediate success that i-mode did. DeNA's auction site met with only moderate success, while its move into mobile social gaming made it one of the mobile game giants in Japan. DeNA founded its own mobile game platform, Mobage Town, which I will return to in the next chapter. Like Matsunaga, Nanba is something of a legend in Japan, with Forbes calling her "one of Japan's most successful women entrepreneurs," and she is one of the only female owners of a Japanese baseball team (DeNA is owner of the Yokohama DeNA BayStars).[55]

Obara has since worked at Matsunaga Mari's home company, Recruit, as well as Google and Rakuten, among other internet companies in Japan. In recent years, he has written two books about platforms and IT business. In his first of these books, titled *The Principles of the IT Business,* Obara gives us a sense of the function of consulting firms like McKinsey, which we can use to think its role in the i-mode project. The role of consulting firms in Japan, Obara writes, used to be to act as "time machines"—relaying information in the form of reports on the newest activities of companies in the United States to the particular Japanese companies that hired them. These reports came before these activities became available in Japan through formal channels such as news organizations or magazines.

As Obara argues, "The reports from these consulting firms have something of a time-machine-like value."[56] The value of the information contained in the report lay in the short time in which it was transmitted. Consulting firm historian Christopher McKenna similarly argues that "consultants acted as the transmitters—or as technological historian Hugh Aitken called them, the 'translators'—of managerial ideas developed in other organizational settings."[57]

Consultants operated as translators, cultural porters of knowledge from one milieu to another. They not only move knowledge across time; they also move it across space and language. According to the business journalist Walter Kiechel, McKinsey aimed to create more value for its customers by creating savings on existing production but also by looking for ways to raise prices or create new products for its clients, particularly from the 1980s onward.[58] This latter activity certainly squares with McKinsey's role in creating a new product—indeed a new product category—in mobile data services for the general consumer, with what became i-mode. We can say, then, that consulting firms are agents of globalization—and in the case of McKinsey's intervention into i-mode's creation, agents of the globalization of mobile technology, strategy, and business practice.

That said, from the manner in which they are often—if somewhat uncharitably—described by those involved from the Docomo side, the McKinsey consultants seem to have been working with a traditional value chain model, beholden perhaps to 1990s wisdom that *contents is king,* particularly in advocating to Docomo that contents should be produced in-house. Matsunaga and Natsuno suggested a different model, one that drew more from the magazine and software industries, both of which were already two-sided markets. Contents should be produced by third parties, they argued, and Docomo would mediate and encourage the sale of these wares. They proposed, in short, the creation of i-mode as a mediation platform. Their contents strategy was the key element of this.

Matsunaga Mari: Contents Editor in Chief

The new media platform business model Docomo adopted had its roots in an old media approach to magazines: pairing advertisements with readers. It was also at the core of Recruit, a recruitment firm that was a consummate "matchmaker"—to use David Evans and Richard Schmalensee's term for platforms.[59] Both come to i-mode via Matsunaga Mari, the "mother of i-mode."[60] Matsunaga was a founding editor of a number of job-seeking magazines published out of the human resources company Recruit, including *Travaille*. Indeed, Natsuno compares the i-mode strategy to that developed by Recruit, where Matsunaga had worked for twenty or so years; the i-mode business model was, he asserts, "created by Recruit." He continues: "Consumers are prepared to pay for Recuit magazines, which are essentially a collection of ads. However, in the case of mobile phones, users don't want to use them for the purpose of acquiring specific information. If we could get the information onto mobile phones, people would start looking at it as a natural extension of using the phone."[61] As Natsuno implies here, Matsunaga did not approach i-mode as a technology—something she had little interest in per se—but rather as a business model based around the use of content as a means of luring and engaging users. Enoki, as we noted already, viewed the development and curation of contents as essential to i-mode's success, for which reason he recruited a magazine editor for the top job at i-mode in the first place. Matsunaga became something like the editor in chief for i-mode content.[62] The term she used to conceptualize the role i-mode would play in the organization of information and its provision was *concierge*. I-mode, she thought, would play a similar service role to that of a hotel concierge—a person who finds information and provides it for the hotel guests. "'Concierge' would eventually become the media concept of i-mode," says Matsunaga.[63] (A very similar model is also behind recent developments in smartphone assistants such as the Google Assistant—which attempts to preemptively organize relevant information for its users.)

The strategy was to curate and present content in the form of menus and submenus that would allow the user to easily navigate the services and contents available. If the concierge was one media concept for i-mode, the convenience store was another. Docomo VP Enoki Kei'ichi compares the i-mode strategy to that of a convenience store, as distinct from a department store. A department store aims to be exhaustive in its merchandise, whereas convenience stores carry a very limited and carefully calibrated array of products. As Enoki explains, "Think about department stores. You don't go there very often. But when you go, you spend a lot of money. Just the opposite is the convenience store. You don't buy lots of things. The variety is limited. But you go there every day. I thought that same concept could be applied to information. Mobile content is inferior to other media in terms of speed and screen size. But in terms of ease of use and convenience, it's far superior."[64] The convenience store is everyday, quotidian, located just around the corner, close-by; it is a place of constant availability, a just-in-time, logistically supported, ready-to-hand logic of consumption that becomes the model of i-mode.

Just how this convenience store–style contents would be gathered and distributed was a point of contention among the i-mode group; the result would in turn have a long-lasting impact on the very conception of how the smartphone business operates. The McKinsey consultants had planned for Docomo to purchase information from various sources, which it would then sell to consumers.[65] Additionally they advised Docomo to charge companies to list their information; banks, airlines, and so on would be charged a fee for appearing on the i-mode service. According to Matsunaga, she and Natsuno were dead set against this model, since it expected too much investment on the part of companies in what was at the time an unproven service.[66] Natsuno also believed that if Docomo competed for content production, other companies would have little motivation to provide good content of their own; it would ultimately "not help sustain service providers' motivation to keep improving their own content."[67] Instead, Matsunaga and Natsuno advocated for Docomo to adopt the role of *mediator, arbiter, and editor* of content rather than the direct provider of it. Moreover, Docomo would not charge companies for offering services through i-mode, since that would discourage them from doing so in the first place—and it needed to launch with as many service providers as possible in order for i-mode to be a good value proposition for users.

In other words, this was a classic moment in what now is recognized to be the platform business model, wherein decisions are made about which side to subsidize and which to charge. As Nick Srnicek points out, "Since platforms have to attract a number of different groups, part of their business is fine-tuning the balance between what is paid, what is not paid, what is subsidized and what is not subsidized."[68] The i-mode team ultimately settled on having outside companies create and offer contents and services, with Docomo taking a 9 percent cut on sites that charged users for their content and acting as a mediator for every "official site." (The official site is the equivalent of an app-store-listed app in today's smartphone milieu; to be recognized as such required undergoing a rigorous screening process by Docomo, and a successful application would benefit the site owner by displaying it in a prominent position on the i-mode portal. I-mode also allowed for unofficial sites, though these would not be listed on the portal or linked from official pages.) This i-mode contents model was set up in such a way so as to ensure that so-called contents providers (companies that sold contents on a subscription basis) or service providers (companies like banks and airlines that offered services free of charge) got on board with the project. The tricky feat of getting content and service providers on board was left to Natsuno, who courted the most serious institutions first, to establish i-mode's credibility: banks. Ultimately, i-mode launched with sixty-seven content and service providers; a mere two years later, the number of providers had grown to over eight hundred.[69]

The simple conception of Docomo as a mediator for the sale of goods and subscriptions through its store had a significant impact on the ecosystem design of smartphones to come—including Apple's and Google's smartphone projects. I-mode created, in short, the precursor to Apple's App Store and Google's Google Play. This revenue-sharing business model saw Docomo take a cut on items or services sold through i-mode (the service charges were often monthly and subscription in nature, with Docomo suggesting that service charges be capped at the reasonable-seeming rate of 300 yen per month, approximately US$3). Noncharging services were offered for free, but as noted above, the lion's share of Docomo's income came from data usage charges. As such, its business model was one very much based around network access fees and service provision—rather than, say, the advertising-driven economic model of Google. Docomo billed for all services, which appeared on every customer's monthly phone bill. At a time when e-commerce—to speak nothing of mobile commerce—was new, paying through one's telecom operator was a reassuring and easy way to make payments for online services for users, who did not have to give their credit card details to unknown companies.

As John Ratliff notes in an early academic study of i-mode, "Only 'official' content providers, accessible through i-modes portal menu, can take advantage of the billing service. Content providers are required to undergo a lengthy application process without clear rules for qualification in order to qualify for 'official' status and thus access to the menu and billing services. This gives NTT DoCoMo great power in its relationship with content providers."[70] In short, Docomo functioned as a gatekeeper that let certain content in while keeping other content out. The difficulty of getting recognized as an official site resulted in the emergence of a secondary market for existing and known contents providers, who often provided consulting services to aspiring contents providers, selling their know-how of how to get listed as official contents with Docomo.[71] Unofficial sites were in a different class; they could be accessed from the device but only if the user knew the web address. Otherwise, these unofficial sites were outside of i-mode's garden walls and effectively invisible to the user. This distinction between official and unofficial sites created a de facto inside and outside of the i-mode system, with those who were officially recognized capable of receiving payment through i-mode. The fastest way for users to access nonofficial sites was to type in their website address directly.[72] That said, Natsuno does note how surprisingly quickly a search engine was built for the unofficial sites—within eleven days of i-mode's launch there was already a search engine released for the unofficial sites.[73] In February 2001, Google announced the launch of a search engine that could search the entire World Wide Web from an i-mode phone—the company's first foray into mobile search, the aim for dominance in what would later become the core motivators of its Android project.[74]

Content Is King, and i-mode Is the Hand of Buddha

Thus far I have emphasized how i-mode was conceived as a business model more than a technology per se. Even so, several choices made around technology were nonetheless key to its success at launch. In the late 1990s into the following decade, there were two main protocols for data services: packet-based services and Wireless Application Protocol (WAP). WAP was a set of protocols designed by a consortium (the WAP Forum, later the Open Mobile Alliance, which included Docomo, even though the company ultimately did not adopt WAP) whereby users choose to connect and disconnect from the network. Much like the dial-up model of the wired internet, with WAP users would be charged for the length of time they use the service, from connection to disconnection. I-mode, by contrast, operated on a packet-based system, whereby users were charged for the number of packets (or data) they used. Early articles on the i-mode phenomenon argued that part of its appeal was its clear fee structure. As reporter Hara Yoshiko writes: "In the conventional phone system, the meter starts ticking at the time of connection. But in packet communications, the fare is leveled according to the number of packets, or volume of data transmitted, not connection time."[75] Being packet-based also meant that the data channel was always open: "While users of other wireless e-mail services must dial in to receive their e-mail and are charged by the minute, i-mode users have continuous e-mail communications and are charged by the byte."[76]

Being packet-based, i-mode was *always on* (whereas WAP-based services such as those offered by i-mode's rival J-Phone would only be activated when turned on, at which point one would have to wait some seconds before the actual connection—as with the dial-up internet). I-mode's connection was constant, and there was no lag in connection time. Email would arrive anytime, notifying users through a sound or vibration, sometimes a blinking light, essentially replacing SMS messages. This left no doubt as to the addictiveness of the phones as messaging devices; email communication became like texting. (A similarly addictive tendency would be seen in the case of BlackBerry's success in North America, albeit almost exclusively among the business community at first, giving rise to the nickname "CrackBerry.") Another key difference between WAP and the packet-based system adopted by i-mode is that the former initially required websites to be written in wireless markup language (WML), a distinct language for use with WAP services.[77] I-mode, by contrast, used a close variant on the HTML standard for web pages, a version known as Compact HTML or cHTML. Given that this language was only a minor modification of standard HTML, it required a minimal amount of additional work from those who wished to build websites for the service. The decision to go with a packet-based service that used cHTML is credited with accelerating the acceptance of i-mode by both users and contents providers, ultimately providing more contents for users.

These decisions around technical standards ultimately bring us back to the emphasis on content. In an interview that further proves Thomas Lamarre's point that new media emerges out of television, Enoki likened the importance of content for i-mode to the model of TV: "It's like TV: Without attractive content, you can't sell TV sets. . . . We considered i-mode a new style of media, so we think content is very important. That's why we decided to use an HTML browser: For them to provide content, it was important for us to use the de facto standard of the Internet."[78] Natsuno, as we saw above, framed this relationship between technological platform and content as a "'positive feedback' cycle," wherein "good content attracts users, which in turn attracts more content, and so on."[79]

In a word: creating a user-friendly, contents-rich platform was their number one concern; one might go so far as to say that for the i-mode creators, *the availability of contents was king.* Crucially, though, they left the creation of contents to other parties, focusing instead on building and securing a content-friendly platform on which to host this content. In one of the many appraisals of i-mode at its peak, Deborah Young calls Docomo "the king of wireless content."[80] If contents seems to have been king, the i-mode system was the platform that made contents possible. Contents was the lure to the platform. So, when it came down to it, *the platform was king* but contents were required for the platform to become so. Content producers make money too, but nowhere near the amount that Docomo raked in from network fees alone. If content is king, the i-mode platform is, in Enoki's terms, the "hand of the Buddha" upon which the world sits.[81] I-mode as the hand of Buddha was, for lack of a better word, a kind of megaplatform that allowed other platforms to develop.

Docomo's decision to forgo the creation of contents on its own went against the tide of the times—when contents acquisition in the form of merger and acquisition deals was rampant—and the advice of the McKinsey consultants, presciently anticipating the age of platform centrism ahead.[82] The rise of Google, Apple, Amazon, and Facebook—the so-called Gang of Four—would prove, if any doubt remained by that point, that platforms were king, not contents. (Though the ascendance of Netflix as content producer and contents platform affirms once again the mutually dependent and coconstitutive relationship between good contents and platform dominance.) In his 2015 book reflecting on the i-mode project and its aftermath, Enoki acknowledges that Docomo and telecoms have lost their pride of place as the hand of Buddha, a position now occupied by current megaplatforms Apple and Google, as well as platform aspirants LINE, Facebook, and Yahoo.[83] I-mode is indeed a preview of this shift from a "content is king" mentality to a "platform is king" mentality, even if it could not maintain its platform kingdom into the Apple/Google smartphone-centric present.

Modeling i-mode's Platform Concept: AOL and Nintendo

I-mode developers had three specific models in mind in their formulation of the service. The first was the editorial model suggested by the magazine format and Recruit's magazines in particular. I-mode would organize and curate content; it would actively manage and organize this content so that the user experience would not be like the anarchic web but rather more like a magazine or a convenience store.

The second model was the U.S. internet portal giant America Online, or AOL. One of the largest internet providers in the United States in the 1990s during the early years of public access to the internet, AOL provided a particular model of curated content. It was a homepage that opened not onto a wild west of what would become the web but rather a "walled garden" with access to particular, selected websites and services. This portal model of a walled garden was an inspiration for i-mode. As Steven P. Bradley and Matthew Sandoval detail in their analysis of the i-mode system:

In developing the business model for i-mode, Natsuno took a look at some of the most successful models on the Internet. One in particular stood out: America Online. AOL had carefully designed a business model in which the company served as a feature-laden portal into the Internet. Proprietary content and features kept many users satisfied to remain within AOL's domain allowing the company to charge advertising and promotional fees to companies that wanted premier status on AOL's home page. . . . AOL's content, e-commerce, and advertising agreements were a major source of inspiration for i-mode.[84]

I-mode hence took as its model not the open web that was in full blossom by its launch but rather an earlier approach to internet services that was much more closed (and, indeed, much closer to the web as it currently is re-enclosed by apps, and by tentacular platforms like Facebook[85]). It took the step of allowing the existence of unofficial sites—some of which would in later years become giants on their own right, with the mobile gaming site Mobage foremost among them—but users were guided toward official sites and, later, preinstalled apps or "appli."[86] This also foreshadows the re-enclosure of the internet that some see starting with the arrival of the smartphone and the appification of the internet.[87] As the journalist Robert Clark remarked about i-mode in a 2000 article: "Some observers have noted that DoCoMo's approach is more like the controlled online service approach of the early 90s rather than the Internet free-for-all of the 21st century. The home page is a portal, as is AOL's."[88] Against the "Internet free-for-all" of the PC world, i-mode created an effective model of the internet that featured contents as a paid service, one that was at least in part based on the prior example of AOL.

Natsuno himself writes glowingly about AOL in a crucial passage on the platform concept partly quoted earlier in this chapter and which I quote in full here.

> We leave the content creation to the service providers who excel at that; Docomo concentrates on our system for collecting fees, our platform, and designing our data warehouse. This is what we might call our platform concept. Docomo's role, on which we focus, is to provide a win-win platform.
>
> America Online (AOL), which grew so rapidly as an Internet service provider, is also based on a similar platform concept. AOL does not create content. It concentrates on designing the platform so that service providers are drawn in droves into the AOL community.
>
> Because i-mode is an Internet-style business, we followed the same business model as AOL. Rather than buying content, we wanted to create a situation in which service providers can make money through Docomo's services. That is, we could work together and both would benefit. The rule we established was: "Docomo makes money from our packet communications charges, while you make money by charging for your services."[89]

Several elements of this passage require our attention. The first is the symbiotic relation between Docomo and its contents providers, which I have already noted above. The second is the framing of both i-mode and AOL as platforms. And the third is the indebtedness in relation to the model of internet service provider suggested by AOL. I-mode was one genesis of the platform concept, in the realm of mobile, but it was also preceded by AOL as another platform pioneer.

AOL at that time had two characteristics in particular that made it an inspiration for i-mode. First, it was a closed system, with a curated front page—the portal—that would bring users to other sites within AOL's "walled garden." As Tim Wu explains in *The Master Switch*, "AOL was, in those early days [in the early 1990s], *the platform*, and, in the lingo, operated as a 'walled garden' for its users."[90] In a *Wired* article from 1996, Frank Rose describes AOL's walled garden in the following terms: "To sign on to AOL is to enter the cyber version of a suburban mall—a carefully modulated, vaguely cutesy-poo environment where the ambience is serenely antiseptic (as long as you don't venture into the wrong chat room late at night) and the impulse to consume is stimulated at every stop."[91] Like AOL, i-mode curated the internet, offering access to a limited number of sites that it carefully controlled within the walled-garden model—an online version of the convenience store, if not quite a mall. It was a closed system but one with a low barrier of entry for users. Convenience and usability were prioritized. And email was its killer app—much as email was the addictive service that got i-mode users hooked too (with a notification sound or vibration going off the moment a new message arrived).[92] But the walled-garden metaphor is deceptive in a very important sense: it is less walled than it is meshed, allowing information to flow in a highly regulated manner, dependent on prior agreements and economic transactions. It is also, given the key relationship between contents, platform, and hardware that we have seen above, multisided and ecosystem-like, implying relative closure but also a certain amenableness to life under a particular set of conditions, in this case set by the platform master, Docomo.

AOL's second notable characteristic from Docomo's perspective was its ability to reduce barriers of access to the internet, making the World Wide Web (or at least an enclosed portion of it) accessible to a much larger number of people than had earlier used it. Like AOL, which had put a floppy disk or CD in everyone's mailbox, Docomo's i-mode provided a means of getting people to access the internet easily, even for those most put off by the technological. I-mode's promotional materials did not even use the word *internet* at all, as one scholar notes: "NTT DoCoMo was careful in its management of consumer expectations, never even mentioning the word 'Internet' in marketing i-mode. Rather, i-mode was advertised as a new kind of wireless data service."[93] I-mode was user-friendly and never intimidating; it provided an easy access ramp onto the internet.

Where i-mode built most on AOL was in its role as contents screener and particularly as financial intermediary, an agent of trust between the consumer and a contents provider or service provider. This financial relationship was fundamental to the i-mode project and entailed a fundamental advance in the platform concept from AOL. As Kokuryō had observed in the early 1990s, the internet was going to require more intermediaries, not fewer. And the greatest intermediary of all—at least for this earlier period of the internet—was to be the Internet Service Provider that gave access to the internet in the first place. Like America Online, albeit with the hardware and software bundled together, and with a payment system perfected, Docomo's i-mode, J-Phone's J-Sky, and au's EZweb each offered easy access to a closed network of services that simply came with one's phone upgrade.[94] These intermediaries offered access to their walled gardens, which they carefully curated by monitoring and judging content pages submitted through a formal submission process. They also took care of billing for these services. The Japanese services in particular became a platform of platforms—that is to say, a platform that allowed other platforms to emerge, proving the point that the layer model of the platform intersects with the mediation model. They created the milieu in which contents services flourished and, in some cases, later became platforms unto themselves. (Where they differed, profoundly, was with regard to content. AOL's failed merger with Time Warner was an attempt to marry platform with content, or internet service with content services. As we have seen, Docomo steered clear of that temptation and succeeded by being a platform that connected users with content supplied by third parties.)

AOL became obsolete when "America" entered the broadband era of internet connectivity and access to the internet was offered by cable companies. By this time, users could go outside AOL's walled garden, yet there was little use for the service once broadband access became the increasingly de facto means of accessing the internet.[95] AOL declined as other means of internet access became more prominent, and as its walled-garden approach became more burdensome than convenient. While i-mode avoided this fate for some time, it was eventually displaced by smartphones—which were in fact in many ways "dumber" than i-mode-equipped phones, not having built-in pay systems, TV reception, or many of the functions users could expect from feature phones in 2008–9, when the smartphone debuted in Japan. Yet despite their limitations, these smartphones could give users access to the whole internet, as well as provide larger screens. Indeed, form factor and touch interface were likely the largest changes from the feature phone, and the same drive for novelty that led customers to adopt i-mode led them to drop it once the paradigm-changing form factor of smartphones came to market. Less than their features—which feature phones by that time far eclipsed—it was the form factor that lead to i-mode's decline, not to mention the cultural values that the slick, new, and desirable iPhone symbolized, at a moment when the virtuous circle of i-mode was stalling. Natsuno attributes this stalling to government regulations briefly introduced in 2007 that prevented carriers from subsidizing phones, leading, Natsuno notes, to a reversal of the virtuous circle between carriers, handset makers, and consumers, and the beginning of a "bad cycle" wherein carriers shifted from demanding phones with new features from handset makers to wanting the same phones at lower price points. That this happened right before the introduction of the iPhone to Japan in 2008 is for Natsuno part of the reason for i-mode's decline.[96] While i-mode's decline has taken place over a long period of time—with many users still preferring the one-hand-holdable phones, especially suitable for the long train commutes many people take every day to get to work—2015 was the last year a new i-mode model was produced; and after June 2017 i-mode phones were no longer available for sale. Interestingly i-mode-style flip-phone form factor phones are still widely available, now running on a modified Android operating system and no longer part of the i-mode system.

With the rise to dominance of smartphones over the early 2010s, the platform mediators became Apple and Google, cutting the telecoms Docomo, au, and J-Phone/SoftBank out of the picture and effectively unifying the world under what were essentially two globally diffused operating systems—outmoding i-mode. Shades of Microsoft's hegemony in the PC world, Android and iOS between them now have almost the entire market of global smartphone sales, with a 99.8 percent market share of smartphone sales in the second quarter of 2017.[97] Together they hold a duopoly of mobile phone shipments. And yet both were clearly inspired by the model of i-mode's success story in Japan.[98]

Before turning to this story of i-mode's legacy and the continuing effects of i-mode, I would like to linger on one other significant influence on i-mode, a third model for its development alongside Recruit-style magazines and AOL: Nintendo. Nintendo may be famous among gamers for generation-defining games like Mario Bros. and Pokémon, but it is famous too among economists and management scholars for the manner in which it set up and mediated a relationship between contents provider and user. (The fact that Nintendo is the publisher, not the developer, of the Pokémon games is a case in point of its multisided platform strategies.) Dwango CEO Kawakami Nobuo has the following to say.

> Apple's App Store and Google's Google Play are representative examples of platforms that screen and evaluate content. The origins of this type of platform are to be found in Japan, with NTT Docomo's i-mode platform that Apple and Google studied and took as a reference point, and, going back even further, the business of Nintendo's video game consoles.[99]

Nintendo was known for its strict control over which companies and games could be released for its consoles. After experiencing incidences of piracy in its first Japanese release of the Nintendo Family Computer (Famicom) console, it released its U.S. analog—the Nintendo Entertainment System (NES)—with an infamous bit of technology that would prevent any efforts to play pirated or unauthorized third-party games on its consoles. This so-called lockout chip was a mechanism embedded in the game cartridge that required a particular kind of "handshake" between the game cartridge and the machine for the game to play.[100]

In addition to tightly controlling who could make games for its consoles, Nintendo also mastered the platform model of two-sided markets as part of this strategy, selling its consoles at cost and making its profits from the games (including by charging independent developers high prices for the chips and other materials needed to make the games in the first place). "What is called the 'Nintendo business model' of the home console is unique," notes Kawakami, who adds: "The i-mode cell phone succeeded by copying this Nintendo model."[101] This Nintendo model that Kawakami speaks of is what we might redescribe as the development of the game industry into a multisided market. Unlike Atari or Coleco, Nintendo carefully coordinated, managed, and controlled those who made games for its console, and the prices for both games and console. It deployed this model for its Japanese console, the Famicom, and engineered this relationship into the very hardware of its North American sister console, NES. In *Invisible Engines,* Evans, Hagiu, and Schmalensee explain Nintendo's platform practice in the following terms.

> Nintendo had actively pursued licensing agreements with third-party game publishers to get a critical mass of games for its new system. However, having witnessed the 1983 U.S. video game market crash, it concluded that in order to succeed, it had to control the quality of games sold for its platform. Accordingly, each NES cartridge contained an authentication chip that was necessary to provide access to the console circuits. Nintendo also kept tight control over the games supplied for its console through its Nintendo "Seal of Quality" policy and, in the interest of quality control, forbade any single developer to publish more than five games every year for the NES.
>
> The authentication chip also allowed Nintendo to charge royalties to third-party game developers, thus converting them from enemies to allies. Nintendo determined the selling price of all games and charged its third-party developers a 20 percent royalty on sales.[102]

Nintendo transformed what was known as the razor-blade strategy of subsidizing the console and profiting from directly selling the games (present
with Atari 2600 as of the late 1970s, and which is ultimately a one-sided
market in the sense that the strategy is an incentive for consumers to buy
the *same company's* other products) to a model of a multisided platform—
wherein Nintendo recruited game designers and earned licensing fees from
them, mediating between multiple players (consumers, software producers).[103] This was enforced by the lockout chip that prevented unauthorized
games from playing on the system, pioneering, in Casey O'Donnell's estimation, DRM by literally "carving . . . in silicon" the ability to restrict access
to nonauthorized games.[104]

To reiterate: i-mode based itself on the Nintendo model, and Apple and
Google in turn based their phones and app stores on the i-mode model.
There is, then, a historical sequence from the VCR–Beta battle that was
at the heart of the network effects literature we encountered in the previous chapter, to Nintendo's and subsequently i-mode's development of
the multisided market in practice, to Apple and Google's redeployment of
this platform concept in the form of their respective smartphone operating systems.[105] This is a sequence that is rarely acknowledged in English-
language literature on the smartphone—with its single-minded focus on
Apple and Google—and yet is common knowledge to Japanese tech writers and tech business operators.

Like Nintendo, i-mode also engineered digital rights management into
its system, preventing users from porting content from their phones to
other phones or computers.[106] Yet in another key way, i-mode differed from
the Nintendo model by design. Docomo did allow unofficial sites to operate. Unauthorized sites could not benefit from Docomo's payment system
for support and were not listed on the main menu of i-mode. But that did
not prevent them from existing, and exist they did, in ever-larger numbers.
Indeed, the tens of thousands of unofficial sites that sprang up soon after
i-mode's launch are also the place from where some of the future platforms
would arise, among them DeNA's mega social gaming site, Mobage Town.
Hence i-mode's greater degree of openness than AOL and Nintendo would
have far-reaching consequences for the Japanese internet ecology, insofar as
it afforded the emergence of future platforms. Natsuno himself emphasizes

this openness of i-mode in distinguishing it from the Nintendo model.[107] Nonetheless, notwithstanding i-mode's relative aperture, its inspiration remained the comparatively more closed Nintendo, which informed the former's inspection-based, closed model of the platform that would in turn be seen in the development of iOS and Android's app stores.[108]

Platform Scheming

Before moving to the discursive and financial effects of i-mode on the contents ecosystem—the topic for the next chapter—it is worth reflecting for a moment on the meaning of platform in this context, and the now current keywords of digital business it inspires. I have noted that the conception of the platform as developed around i-mode is one that figures as both a basis for activities and a model of mediation. It is a manner of manipulating multiple markets into a virtuous circle—a self-sustaining *ecosystem* of content, device makers, users, and network operators.

The emphasis in this account on business models as a way of narrating internet histories returns us to an important element of the i-mode story: ultimately, this is a success story of a concept, or system, more than it is of a technology per se. I-mode, as many journalistic accounts around the year 2000 have it, was a test bed for mobile commerce. "I-mode has made Japan the incubator for mobile commerce," says one journalist.[109] "Japan has become the test-bed for the world's mobile internet future and the m-commerce incubator by default," writes another.[110] If it was indeed a future of mobile commerce, it was such because it constituted one of the first successful implementations of a commerce-based mobile internet system. It is *transformative* insofar as it creates a new mode of mediation between multiple partners, pioneering the emergence of the mobile internet as a fee-driven platform. That said, once in place, the form of mediation is closer to the model of the intermediary discussed in the previous chapter: a nontransformative, repetitively and habitually economic, monetary form of exchange put in place by i-mode. This repetitive exchange was also the source of value creation in the form of the contents market, and, more important for Docomo, it was a mechanism for hooking users on internet services, boosting its profits through massive fees in data costs users incurred.

As such, i-mode (as well as its competing systems in Japan) was a technologically enabled mechanism for organizing people and behaviors. While i-mode could be narrated as the development of a set of technologies, it can be more accurately described as an implementation of a set of techniques; it is a narrative of techniques, of *mechanisms (shikumi).* Indeed, it is as a mechanism that Natsuno describes the ingenuity of i-mode (and Apple's iTunes-iPod-iPhone hardware-services ecosystem subsequently). Japanese companies must shed their obsession with "making things" *(monozukuri),* Natsuno admonishes, and should instead be preoccupied with "making mechanisms" *(shikakezukuri).*[111] Another translation for the term for mechanism *(shikumi)* that emphasizes its more sinister dimension is *scheme,* as in "the byzantine scheme of capital accumulation" that Sumanth Gopinath describes so vividly in the case of the mobile ringtone industry, a description that fits so well with the multiple contents industries—ringtones, wallpapers, surfing news, and so on—propped up by i-mode.[112] It is a mechanism for payment and a scheme for mobile addiction whose ramifications are so proximate to us today and are being explored by psychologists, technologists, filmmakers, and artists alike. And it is a mechanism that sediments experientially as habit—along with addiction (and perhaps functionally indistinguishable from it), the most important concept for explaining that most intimate relationship with technology that the always connected mobile phone brings about. As Wendy Hui Kyong Chun points out: "Habit is becoming addiction."[113]

This mechanism is itself a form of mediation, urging us to read Natsuno's mantra as a shift from making things to making mediations. The rise of the platform economy marks a further intensification of a shift from making an object to be sold to making an ecosystem of interconnected markets, consumers, and objects. Ultimately, i-mode shows the success of the mediation platform as outlined in the conclusion to the previous chapter: it is a mechanism that expands the management of a company to the management of the social body (via the totality of users of i-mode) in toto. Through an expansion of the mediating middle, i-mode exerts control over its users and the social body in general via its platform logic. And this is the real feat of i-mode: generalizing its control over the social body, in a manner that could be repeated by other platform entities, elsewhere.

The smartphones of Apple and Android, and the similar attempts by BlackBerry and Windows Phone (the latter two often framed as "losers" of the ecosystem wars), adopted and expanded these mechanisms and their managerial control. The smartphone allowed access to the wider internet (i-mode required specially formatted HTML pages), and in a hardware cycle dictated by novelty, the large screens of the smartphones ultimately won consumers' hearts. But smartphones are fully dependent on the i-mode paradigm as a model; except rather than the telecom, it is Apple or Google that became the clearinghouse for payments processing. They became the operators of the transactional platforms now under the form of app stores, music stores, news aggregators, book readers, and so forth.

Even as the Japanese telecoms lost their place as the *platform of platforms,* megaplatform, or the palm of Buddha, as Enoki memorably put it, they were supplanted by other platforms that took the lessons of i-mode and applied these lessons within a different configuration. This time the giants of Silicon Valley, Apple and Google, were the palms of Buddha, the platform owners and managers. Not being wedded to the infrastructure of mobility that telecoms operated, they had different sets of priorities and allowed different kinds of relationships (for example, they allowed the proliferation of chat apps, something that some believe could not have happened under the reign of the telecoms). And yet they were also clearly a continuation of the mobile strategies developed in Japan around i-mode and other mobile phones—interfaces to a world of contents and services, and mechanisms by which users were engaged and even addicted, making the mobile phone a core part of their everyday lives. This includes the development of mobile payment services, which were first introduced in Japan in 2004, through the Felica standard. Docomo took a step further in the direction of financial services, launching its own credit card—the DCMX card—in 2006. (This moment could be considered "peak platform" for Docomo insofar as the mobile phone technological platform intersects fully with the mediatory credit card platforms described in the work of Kokuryō, as well as Rochet and Tirole in the previous chapter.)

Following their introduction in 2008, Google and Apple smartphones ultimately ate into the "feature phone" market share in Japan, just as they eventually took the place of BlackBerry elsewhere. But they also took the lead from i-mode, building themselves their own ecosystems and contents mediation and situating themselves as the center of the users' mobile universe. Data gathering became increasingly important to this in a way that was not yet the case with the Japanese phones. Siri and the Google Assistant take the concept of the concierge to the next level, providing help before it is asked for, anticipating the needs of its user based on past (data-gathered) experience, and in turn driving other services based on this data—yet this is also very much an extension of transformations i-mode pioneered (albeit with some inspiration from Nintendo, AOL, and platform theory). First in i-mode then in the iPhone we witness the practical extension of platform management theory as the total mediation of life. As a result, the platform as scheme for total mediation is more thoroughly and more globally in place than ever before. The i-mode lesson was learned.

CHAPTER FIVE

Platforms after i-mode

Dwango's Niconico Video

The Docomo, au, and J-Phone/Vodaphone/SoftBank proto-smartphone mobile internet systems—i-mode, EZweb, and J-Sky, respectively— were the platforms that created the conditions for the blossoming market in contents. Due in no small part to the emphasis on contents as commodity within the i-mode and other mobile internet systems in Japan, contents gained monetary value and the discursive weight tracked in chapter 1.[1] The discursive inflation of contents to something of a master language and keyword of the early twenty-first century in Japan can only be understood in light of the convergence of several factors, which include: (1) the emergence of software as a commodity, particularly in the 1980s, and the concomitant realization on the part of hardware manufacturers like Fujitsū that "software"—and later, contents, as a medium-agnostic term that describes media forms —would be a growth area; (2) the soft power "Cool Japan" initiatives of the Japanese government around the global popularity and globally distributed anime, manga, and game contents (or the media mix of these); and, perhaps, most decisively for this chapter, (3) the creation of a market for digital contents through the structures put in place by i-mode, EZweb, and J-Sky. Thanks to the latter, contents companies began raking in money from subscription-based services, later launching splashy IPOs. There was a concomitant inflation of stock price and discursive value that accrued to the word *contents* that was dependent on the strength of mobile platforms as mechanisms for user engagement and contents monetization.

I-mode, EZweb, and J-Sky were also, and perhaps even more importantly, the platforms that allowed other platforms to emerge. If one of the definitions of platform is that it is a base from which other systems emerge (or upon which others can be built), i-mode and its peers function as a base or megaplatform from which other platforms spring, often from within their contents providers. These newer platforms get their start as contents providers and then later become platforms themselves. Unsurprisingly, these fall into the category of *contents platforms,* being contents providers who then become platforms for other contents providers, including for user-created content. Yet they are also later evolve into structures of mediation for third-party contents, forming the multisided market platforms examined in the previous chapters. Three of the most significant platforms for their impact on the mobile media milieu and Japanese internet culture more widely are DeNA's Mobage Town (now Mobage), GREE, and Dwango's Niconico Video (Niconico dōga). Mobage Town and GREE are known as social networking sites (SNS) that focus around social games in particular; despite their claim to being SNSs, their principal function is as platforms for social games.[2] More recently, since the rise of smartphones, they have become game publishers. Niconico Video is a video uploading and commenting site that also fosters an active creating and commenting community. These examples—and the Dwango case in particular—reveal the degree of integration of platforms and contents producers that come, in part, out of the beginnings of these three within the mobile contents provider milieu. The process of transformation that sees these three move from contents providers to platforms also sheds light on the process of *platformization,* and the resultant hybrids that emerge. The conditions for platformization will be the subject of this chapter, and the integrated nature of platforms and contents the subject of the second section of this chapter, focusing on Niconico Video and the particular conjunction between contents and platform that it enables.

This chapter begins by analyzing the contents bubble that i-mode and its comparable platforms gave rise to. It then shifts to a mapping of the platformization of contents providers, which begin to challenge the platform dominance of the telecoms. DeNA's Mobage Town, GREE, and Dwango's Niconico Video can appear to be halo effects of i-mode, J-Sky, and EZweb—

beneficiaries of the megaplatforms from which they sprang. Yet it was also thanks to certain crises and transformations in how these megaplatforms functioned that these three novel platforms got their start. The second section of this chapter then turns to a close analysis of Dwango's Niconico Video platform, in order to show both how i-mode provides the basis for future platform development, as well as to turn to the cultural politics of platforms, this time attending to their interface, in order to address how platforms intersect with contents. I will ultimately argue that Dwango's origins and identity as a contents provider fundamentally shapes the type of platform it produces. This chapter then addresses the conceptual framework of platform imperialism—brought to bear on global dominating powers of Apple, Google, Amazon, Facebook, and Netflix—and explores its usefulness for thinking about an alternative aesthetics and a specific media ecology. Subsequently, I note some of the nationalist dangers of the platform imperialist framework, and the ease with which it plays into a narrative of a resurgent Japan—wherein contents creation becomes the newest platform for regional (or global) expansion. We then see how Niconico Video inspires in some a vision of a creative industrial logic that has parallels to the managerial impact of Toyota's production strategies (known as Toyotism) in a previous era. This chapter concludes with a consideration of platform capitalism as a keyword, building on this chapter's findings to determine whether it is a periodizing entity (like Fordism or post-Fordism) or rather a subcategory of post-Fordist modes of accumulation.

From Contents Bubble to the Platformization of Contents Providers

One of the most important effects of i-mode was its generalization of the sense that one should pay for contents on the internet. It created the mechanism for payment and naturalized the exchange of money for these contents. As Natsuno writes in 2002, "While hopes for a market for fee-based digital content ran far ahead of reality on the wired Internet, a fee-based market did blossom with the rise of i-mode."[3] Indeed, far from simply creating a market for fee-based contents for cell phones, i-mode also created and ingrained the very culture of purchasing digital contents. It made the exchange of money for contents a habit.

Noriko Manabe, in her work on the mobile ringtone market in Japan, gives the following description of the market for contents circa 2005.

> The Japanese mobile internet was estimated to be six times the size of that in the United States in 2005. The market for paid cellular phone content in Japan was 315 billion yen ($2.6 billion) in 2005, half of which came from music. Specifically, 105 billion yen came from polyphonic ringtones *(chaku-mero)*, 46 billion yen from mastertones (ringtones sampled from the master recording, or *chaku-uta*), and 10 billion yen from full-track downloads (recording of the entire song, or *chaku-uta full*). As of 2006, 55 percent of KDDI's content revenues came from music, with manga and e-books growing in revenue; for NTT Docomo, which was late in offering *chaku-uta full*, 20 percent of web access was for music, 24 percent for videogames, and 27 percent for other entertainment applications, such as sports and gambling simulations.[4]

As Manabe details, there was a huge market for paid-for contents in Japan, supported by the mobile ecosystem. The mobile market for paid contents far exceeded the market for PC-based contents, once again showing that the former was configured as a milieu in which one paid for informational goods, unlike the PC market. A survey of the Japanese market for internet-distributed music from the second quarter of 2006 indicated that 89.7 percent was sold via mobile, amounting to 22 billion yen (approximately US$200 million) while only 10 percent of all sales were via PC.[5] In other words, while record companies in North America and Europe were bemoaning the loss of income to the piratical activities of users of Napster or peer-to-peer file sharing, mobile users in Japan were handing over large sums of money just to use ringtones, and eventually to gain access to full songs.

Contents, we may recall from chapter 1, always means packaged IP, or information packaged for purchase. How information or media was formatted as a purchasable commodity is a process highly dependent on the Japanese mobile internet. To appreciate the full importance of i-mode as a tool for formatting the internet as a site of commercial exchange and of *contents for purchase,* it is worth looking at the following exchange between Kawakami Nobuo—former CEO of Kadokawa, CEO of Dwango, founder of Niconico Video, and one of i-mode's earliest contents providers—and Kurita Shigetaka, one of the main programmers at Docomo during the i-mode project, responsible for, among other things, the push-style email program, and best known as the inventor of emoji. The exchange took place in November 2013, and was first hosted and edited by the online site 4Gamer.net (whose comments are registered as 4G) before being published in book form. Given how much bearing this exchange has on the arguments made in *The Platform Economy,* I quote this dialogue at length.

> Kawakami: [I-mode] is said to have created the culture of actually paying for net services.
>
> Kurita: Yes, that's true. One of the main particularities of the Japanese cell phone contents market is that there's the atmosphere that it's natural to pay money for contents. It's hard to find other countries where people would actually pay for things such as the news, or even the weather.
>
> Kawakami: One of the reasons for the success of Niconico was, I think, thanks to i-mode. Since most of its users were part of the generation who were used to i-mode, many people became Premium members of Niconico. [Premium membership costs around US$5/month and comes with benefits such as faster loading speeds and higher resolution video.]
>
> 4G: I see.
>
> Kurita: Now that you say this, I remember having a similar conversation with you along these lines before you started Niconico Video. You had mentioned that while the PC Internet was predominantly organized around the advertising model, the cell phone Internet had established the paid model *[kakin moderu].* At the time, I remember you saying something to the effect that "The PC side of things should also be able to establish the paid model. I'm thinking of bringing cell phone culture's 'paying for things' model to the PC." And indeed, that's what you did.
>
> Kawakami: Yes, we were debating whether the not-paying-money PC side would be the future, or the cell phone model of paying [for access to contents] would be the future.

4G: And what was the conclusion of this debate?

KAWAKAMI: Basically, my feeling was that the cell phone [paying model] would be the future, but that in fact in the end we would fall into some middle ground between the two models.

KURITA: In recent years, the cell phone model of an Internet contents market has at long last developed even outside Japan. But the present state of things seems to be controlled by Google and Apple, which has given people on the contents side of things a sense of crisis.

KAWAKAMI: That's so true. A lot of people seem to think that "Google and Apple built the contents market for smartphones." But when you look at it from the position of the Japanese cell phone contents market, they didn't build the market but rather destroyed the market.[6]

To summarize briefly, the main points of this dialogue include: (1) the distinction between the ad-based revenue model of the PC internet versus the willingness to pay for contents established with i-mode and the cell phone internet; (2) the bleed of the cell phone internet over to the PC internet, as people become willing to pay for contents generally (which sees the reformatting of the PC internet along the cell phone model); (3) the model of paying for services and contents formerly relatively unique to Japan has been generalized globally—if we can use that fraught term—by Google and Apple. Indeed, even paying for news has become more commonplace, as newspapers put up paywalls and depend increasingly on digital subscribers.[7]

The closing remark about Google and Apple destroying the contents market is a fascinating one that I cannot deal with at length here. Briefly, though, the actual numbers point to the exact opposite: since the arrival of smartphones, the total market for contents has continued expanding. In 2006 the mobile contents market for feature phones in Japan was 366 billion yen; 472 billion yen in 2007; 483 billion yen in 2008; and 552 billion yen in 2009. In 2010 the market for feature phone contents was 600 billion yen, a number that shrank over the next few years as revenue shifted to the smartphone (in 2010 the smartphone accounted for only 1 percent of the contents market in Japan, however).[8] During the 2010s, the total market for contents across both feature phone and smartphone expanded steadily, reaching 734 billion yen in 2011, 851 billion yen in 2012, and 1.7 trillion yen in 2013.[9] In other words, contrary to Kawakami's suggestion, total spending on mobile contents has significantly increased.[10] That said, the arrival of the smartphone has destroyed several important *segments* of the contents market and the players dependent on them, most notably the market for ringtones, since with the smartphone one could set a song as a ringtone; it additionally reduced support for some song download subscriptions and some feature phone games—all market segments Kawakami's companies were heavily involved in.[11] The arrival of iPhone and Android phones made people's willingness to pay for certain contents such as ringtones evaporate rather quickly. In the game world, the smartphone also continued a shift from a subscription-based model of social games to a freemium model first pioneered by Japanese companies fostered by Mobage Town and other emergent platforms during the feature phone era. Hence, yes, there was a dramatic shift in market organization that destroyed some formerly vital parts of the market, one of the many disruptions brought about by the shift from feature phones (and the telecommunications companies as megaplatforms) to smartphones (and iOS/Android as megaplatforms). This also coincided with an era of increased competition among contents providers, mediated by Apple and Google, companies that were, in consultant Obara Kazuhiro's estimation, far less concerned for the well-being of the contents providers. I-mode had what Obara calls a "healthy protectionism model" of contents care, compared with Google's and Apple's "survival of the fittest" attitude toward app makers within the smartphone market.[12]

And yet despite the transformations, significant elements of the feature phone model of mobile contents persist. Here we may turn to Sumanth Gopinath's question about the demise of the ringtone: "How will the ringtone, once having disappeared, haunt *us* in global capitalist modernity?"[13] This we may rephrase slightly, to encompass the same era of the mobile ringtone market's demise: How will the internet-connected feature phone, once having disappeared, haunt us in global platform capitalist modernity? The answer is this: through the persistence of the business model of platforms, and through the *cultural habit and technological mechanisms of payment on platforms,* installed in users via i-mode by way of the ringtone economy and the feature phone model of the "walled-garden" web, and instilled globally through Apple and Google. Despite i-mode's disappearance, the platform capitalist logic first developed in i-mode continues to haunt contemporary cultural-technological practices and mechanisms of paying for smartphone-delivered contents.[14]

"I-mode has made Japan the incubator for mobile commerce," wrote Irene Kunii in a *Business Week* article in 2000; it continues to ring true today, even if we must transpose it to the past tense.[15] I-mode *was* an incubator, and a training ground, a form of social conditioning that created an understanding of the phone as a transactional space and the user as a consumer of paid contents. It became a site from which people would learn how to consume contents via the internet, mediated by monetary exchange, and facilitated by the ease of financial transactions mediated by megaplatforms like i-mode and later Apple. I-mode was the global prototype for paid, platform-mediated, internet-distributed contents consumption.[16] Its continuing importance is such that I term the consequences of it *the i-mode effect*.

The Golden Years of Contents and the Japanese Contents Bubble

Docomo's emphasis on the importance of contents for i-mode offers one additional motivating factor for the contents discourse bubble mapped out in chapter 1. Thanks to the market for digital contents delivered via i-mode and its competitors, and the market for contents mediators that it created, contents appeared to be a lucrative field at the very moment the media industries were suffering at the hands of digital piracy elsewhere in the world. I-mode and its comparable services allowed contents to be monetized. Mobile providers were the gatekeepers to and mediators of this lucrative contents market, and a number of the contents providers acted as service points to help aspiring contents providers access this market.

The earliest contents providers were often smaller start-ups. With the exception of Bandai—the massive entertainment and media firm was an early adopter—many contents providers were smaller companies that grew thanks to the market i-mode provided. I-mode was something like a basin for the creation and incubation of start-ups, aiding them to brainstorm possible contents products for the platform and helping them get through the rigorous screening process before they could be listed as an official site (as chapter 4 detailed, only official sites would be promoted on the i-mode portal and could run payments through the telecoms' payments processing systems).

Some of the early contents providers include companies such as Dwango, Cybird, Index, and MTI. In some cases, such as Dwango, these companies had already been established by the start of i-mode, yet often in other computer or mobile-related businesses. Others, such as Cybird, were start-ups created to respond to the i-mode venture, looking to capitalize on the opportunity to make and sell contents for the mobile phone.[17] I-mode spurred their businesses to success, and many of these companies subsequently went public with valuations supported by their income derived from the mobile sector. As Natsuno enthuses:

> Lately, we have seen a gold rush of i-mode-related startup companies making initial public offerings (IPOs) of their stock. The first surge of IPOs by such companies came in 2001: Livin' on the Edge, Index, Access, Cybird. Having done much to popularize i-mode itself, these startups were listing their shares on the over-the-counter market, on the Tokyo Stock Exchange's Mother's (a market for high growth and emerging stocks), or on JASDAQ (the Osaka-based share-trading market for smaller companies' stocks).[18]

Natsuno lists a number of the companies that went public thanks to the i-mode effect (see Figure 6). Indeed, one witnesses a generalized rise in IPOs on Mothers and JASDAQ during the height of the i-mode years (see Figure 7). Some of these companies subsequently used the income derived from their valuations to go on buying sprees overseas, investing in nascent mobile markets in the belief that the mobile revolution underway in Japan would be mere months away overseas, and there for the taking. In most cases, this did not turn out to be the case, and many subsequently shed their overseas purchases at a loss. For instance, Cybird purchased Montreal mobile entertainment company Airborne in 2005 for $90 million, and then sold it back to its original owners "for way, way less," according to Andy Nulman, one of the original owners.[19]

Many of these companies Natsuno lists were involved in one or more of the areas for mobile commerce at the time: ringtones and eventually full

Company	Area of business	Date listed	Market
Livin' on the Edge	Internet-related system development and content development	April 2000	TSE Mother's
ValueClick Japan	Online advertising	May 2000	TSE Mother's
PA Co., Ltd.	Online want ads and employment service	July 2000	TSE Mother's
Magclick Inc.	Online advertising	September 2000	JASDAQ
Mediaseek Inc.	Content development support	November 2000	TSE Mother's
Cybird Co., Ltd.	Content development support	December 2000	JASDAQ
Access Co., Ltd.	Development of embedded software	February 2001	TSE Mother's
Nihon Enterprise Co., Ltd.	Content development support	February 2001	JASDAQ
Open Loop, Inc.	Security technology development	March 2001	JASDAQ
Index Corporation	Content development support	March 2001	JASDAQ
Faith, Inc.	Music content creation	March 2001	JASDAQ
Zentek Technology Japan, Inc.	Development of embedded software	August 2001	JASDAQ
BeMap, Inc.	Content development support	January 2002	JASDAQ
GignoSystem Japan, Inc. (formerly PhotoNet Japan, Inc.)	Distribution of digital images	March 2002	JASDAQ

Figure 6. Initial Public Offerings (IPOs) from mobile internet companies related to i-mode. Natsuno Takeshi, *The i-mode Wireless Ecosystem*, 55.

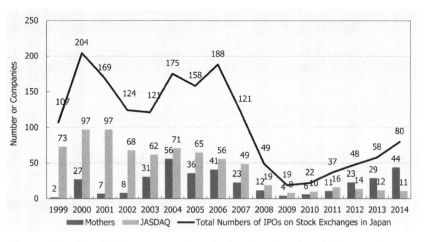

Figure 7. Number of IPOs per year on the tech-related stock markets, Mothers, and JASDAQ around i-mode's peak. Hayase Goto, "The Current Condition and Framework of IPOs on Junior Markets in Japan," 2.

song downloads, mobile games, wallpapers, as well as the secondary market for helping other contents developers make it through the i-mode screening process. Despite their apparent simplicity, there was a huge market for both ringtones and wallpapers, and companies made good coin on the impulse toward mobile personalization that drove those markets.[20] But early contents for mobile phones also included weather sites, surfing information sites that provided the surfing conditions all around the country (such as the very successful Namidensetsu or Surf Legend, by Cybird), and, of course, mobile games, which would later become one of the most important types of paid-for contents. Successful companies such as Cybird also became brokers between aspiring i-mode contents providers and Docomo, functioning as intermediaries themselves. They sold the know-how of getting listed as official contents with Docomo.[21] Contents providers also became consultants to emerging mobile phone contents markets. Cybird, for instance, provided advice to Asian and European telecoms on how to develop contents during the early twenty-first century.[22]

That these contents were provided on a subscription basis assured the companies involved a continuous revenue stream and presages the shift from ownership to renting that marks this current stage of capitalism (in which the sharing economy should be seen as an extension of the renting economy—or what Jeremy Rifkin at the turn of the millennium called the "age of access").[23] One did not simply buy a ringtone; one bought access to any number of ringtones that changed each month, a selection from which one could pick one's current ringtone. The same applied for games. In what was an early example of the now prominent business model of having constant game updates, one purchased access to the game, which would change over time, perhaps responding to atmospheric circumstances or times or day, or weather patterns—as was the case of Dwango's fishing game, Tsuribaka kibun (Crazy about Fishing), for instance.[24]

The glory days of these contents companies is part of the i-mode effect—the spillover effect of the i-mode platform on the contents market and beyond. Moreover, this contents bubble coincided with the heyday of the Japanese contents discourse, arriving at the same time as the Japanese government was taking notice of the overseas successes of Japanese anime, manga, and games, and during which time the government was formulating its first Cool Japan strategies. Together, these constituted what we might call the first golden era of contents.

Telecom Transformations and the Platformization of Contents

In Japan, the introduction of the iPhone in 2008 (initially under exclusive contract with SoftBank) and Android in 2009 (deployed by au and Docomo as they sought to catch up and offer a comparable smartphone model to compete) displaced the carriers' platforms; i-mode, EZweb, and J-Sky did not operate on the new smartphones. Instead, purchases and subscriptions were rerouted through Apple's App Store and Google's Google Play. These replaced the carriers as the mediating platforms for mobile commerce and payment. With Android and iOS, the carriers were cut out of the picture, reduced from being platform operators to being mere "dumb pipes" for the carriage of data and phone connections. Apple and Google thereby overtook the national telecommunication companies' platforms to become the mediators-in-chief and the conduits for all mobile contents—a transformation bound up with the growing influence of these tech companies globally. It is in this context that communications studies scholar Dal Yong Jin has suggested the necessity of resuscitating the cultural imperialism framework, this time as "platform imperialism," a conceptual frame I return to below.

As we saw from Kawakami's comments above, the entry of the smartphone into the Japanese market is also often blamed (or credited) for the decline of the telecoms and their loss of platform hegemony in the mobile space. As with all things the situation is somewhat more complex. The telecoms rode high on the i-mode and other mobile services. As I noted in the previous chapter, they took a 9 percent fee from all transactions and controlled the official contents on their respective platforms. They also had full control over handset makers, which ensured that handset makers were at their beck and call, and produced phones to the telecoms' specifications.

This also had the secondary result of the phones being incompatible with international standards, giving rise to the "Galapagos" discourse around mobile media circa 2006–7, which implied the closure of the mobile market to outside influence and forces, much like the Galapagos Islands about which Darwin wrote.[25] While Galapagos discourse started as a vein of praise for Japan's unique mobile ecosystem, it quickly transformed—principally under the effects of a 2007 book by the Nomura Research Institute think tank—into a critique of Japan's closure to the outside world, and a call for deregulation, greater capital inflows, and more openness to the global. (The translated title of this book is *Japan in 2015: Toward a New Era of "Opening Japan to the World" [kaikoku]*, referencing the Meiji era in the nineteenth century when Western powers forced Japan to open its borders.) As such, the Galapagos discourse is fascinating for the ways it conflates the aperture or closure of the mobile feature phone as walled-garden enclosure *and* special ecosystem unique to Japan with the supposed closure of Japanese markets to the world in toto. Indeed, the mildly derogatory term *Galapagos cell phone* or *"gara-kei"* (a contracted version of "Galapagos *keitai* / cell phone") came to denote the feature phone celebrated in times past. This discourse signals both the crucial place the mobile market of Japan's cell phones had in the national imaginary and the incredibly powerful hold the mobile industry had on Japan's entire industrial formation during the early twenty-first century.

This narrative about the entry of the smartphone and corresponding decline of the telecoms is—like most narratives—true only in part. The telecoms did lose their foothold in the platform sphere with the arrival of the smartphone, and in this sense their importance as platforms diminished.[26] Yet the problem with the "from platform to dumb pipes" narrative is that it neglects the longer process of platformization of contents providers—an important development in the mobile realm around 2006, and a turning point in the telecoms' longer process of decline as megaplatform masters. The platformization of contents is also an important side effect of the mobile internet phenomenon. It remains part of the larger i-mode effect, described above, whereby the mobile internet provided the breeding grounds for the next generation of Japanese platforms, which begin to displace the telecoms in small yet important ways as the brokers and distributors of contents. It also explains the prominent role Japanese game publishers play in the current app-based mobile market. Many of the freemium model social gaming companies that lead the world globally in in-game revenues for their apps come from Japan and emerge out of the platformization of contents.

Allow me to explain what I mean by this term *platformization*. Anne Helmond gives a definition to the term in her influential article, "The Platformization of the Web: Making Web Data Platform Ready." There she examines, in her words, "Facebook's development as a platform by situating it within the transformation of social network sites into social media platforms." "Platformization," she continues, "entails the extension of social media platforms into the rest of the web and their drive to make external web data 'platform ready.'"[27] For Helmond, the platformization of the web is a process whereby social media platforms—what I have called "contents platforms"—extend their tentacles into the rest of the web, following users in their browsing activities, using API plug-ins to permit Facebook commentary on far-flung websites, allowing Facebook the ability to sell users ads about things they were browsing, and generally connecting the web at large back into the closed Facebook platform. David Nieborg and Thomas Poell develop the concept of platformization in another direction, shifting our attention to the effects of platforms on the form and mode of production of contents. This is what they term the "platformization of cultural production," which is to say, the ways that content production is itself affected in its form by the platforms they aim to serve.[28]

While these two uses of platformization are helpful, in this chapter I use the term *platformization* in a more specific if limited sense, referring to the process of becoming a platform, and especially to the transformation of contents providers into a combination of (1) contents platforms and (2) mediation platforms. In so doing, contents providers themselves become the basis for user- or third-party-generated contents, turning into a market for the selling or providing of third-party contents.

There were three conditions for the platformization of contents. First, there was a shift to how data fees were charged and collected. We should recall the degree to which the telecoms' earnings were dependent on data charges from the beginning, and therefore central to their business practice and contents strategies. Good, abundant content was a hugely important driver for i-mode adoption and usage, but regarding their bottom line it was most important to Docomo and the other telecoms as a source of data fees. Ultimately, they made far, far more money on data fees than on the 9 percent cut they took from the contents providers for monthly subscriptions (Natsuno suggests Docomo made approximately 70 billion yen [US$600 million] per year in data charges versus around 3 billion yen [US$27 million] in contents charges).[29] And yet this started to change in 2004, when the carriers (with au leading the charge in late 2003) introduced flat-rate data packages, allowing users to use unlimited data each month, paying a flat rate once a certain level of data consumption had been reached.[30] At around 4,200 yen (US$40) per month, the flat packet rate was not cheap, but it went some way toward the transformation of carriers into flat-rate data transmission services. It also allowed more data-heavy services to emerge, such as portal-based browser games—namely those from GREE and DeNA's Mobage Town.

The second such condition was the beginning of an era of mobile number portability beginning in October 2006, mandated by the Japanese government, which was intent on offering consumers more flexibility in their mobile carrier. Under the new guidelines, mobile phone subscribers could for the first time retain their existing number when they migrated to another carrier. Removing the barrier of having to change one's phone number when changing carriers encouraged user mobility from one platform and carrier to another. For contents providers this change created an incentive to start offering one's content services across telecom platforms, thereby "altering the mobile contents environment," as technology writer Ishino Junya notes in a book on Mobage Town.[31] One of the responses to this was for contents providers to set up portals and user-specific logins and IDs that would allow customers to use their sites, access their existing subscriptions, and download their contents across phones and carriers. In other words, contents providers themselves helped consumers port their services from one phone and carrier to another. Dwango was a pioneer in this, setting up a portal site that could be accessed from the phones of all three carriers even before the era of number portability.[32] When Dwango established dowango.jp as a means of allowing ease of transposition of content from one device to another, it marked the beginning of Dwango's platformization, becoming carrier-agnostic, if not completely carrier-independent.

The third condition for the platformization of contents was the coexistence of official and unofficial sites. For all of their ability to funnel attention and subscription money into themselves, i-mode, au, and J-Phone were not completely closed platforms, as we saw. There were official sites, which they promoted, but they also allowed the existence of unofficial sites, which could be accessed by users. Unofficial sites could not use the telecoms' payment mechanisms, but they also did not have to suffer through their strict screening process and operation guidelines. Some of these unofficial sites themselves became platforms that mediated access to third-party services, including social games. Once again GREE and Mobage Town are the best examples of sites that could not have passed the screening process (mainly since they relied on income from in-game purchases, which Docomo and other carriers forbade in official sites), but they flourished as unofficial sites.

I-mode and its like services by au and J-Phone set up the conditions for the creation of contents as a product but also and secondarily the conditions for these contents providers to become platforms in their own right. Ishino Junya in particular notes that the above three conditions were the enabling factors in the emergence of online social gaming portal sites such as Mobage Town and GREE; while he calls them *portals,* we would do better to think of them as platforms (I will return to this relation between portal and platform in the conclusion).[33] Natsuno Takeshi himself notes the importance of DeNA and GREE, describing them as "platforms on platforms," emerging at a moment when the carriers had lost motivation and direction as contents brokers.[34]

DeNA has particular significance for this book insofar as it was co-founded by Namba Tomoko, the lead consultant on the McKinsey team involved in the creation of i-mode. DeNA began by launching a number of web-based auction sites, before eventually starting a mobile auction site, Bidders, followed by its most successful venture, the Mobage Town portal (or platform), which hosted third-party games, as well as DeNA-made games. As its Wikipedia page recounts, Mobage "is a portal and social network for games" that provides "a platform for developers to deploy their games so that they can be discovered and shared by mobile game users"—of whom there were approximately 30 million at its peak. Mobage is a social media site as well as a distribution platform that became one of the biggest social gaming companies in Japan.[35] Both Mobage and GREE (the latter founded in 2004, launching its mobile services in 2006) pioneered freemium game ventures in Japan. To some—especially those making games for sale or subscription on telecom platforms—they were also two of the major players that undermined the sale and subscription contents market for games circa 2005. Notwithstanding the habituation to pay for contents, the impetus to subscribe to games was diminished once one could access them for free.[36] Mobage and GREE's rein was in turn challenged by the rise of smartphones and the corresponding shift from a browser-based market, in which they acted as portals, to an app-based market mediated by Google's and Apple's app stores.[37] In his careful history of Japanese social games, *The Third Wave of Japanese Games*, Atsuyo Nakayama offers a useful graph that maps the transitions between eras of platform dominance— from hardware platforms (game consoles), to the telecoms as platforms, to the mobile game platforms, to the smartphone platforms (see Figure 8).

The pattern can be described in the following terms: Docomo's i-mode is the platform for the development of third-party contents providers. Other companies develop services that are run through and mediated by the Docomo i-mode platform and its store or web page. Subsequently, a few of these contents providers break away or form their own sites or platforms. These are web-based services formatted for mobile phones, in many cases,

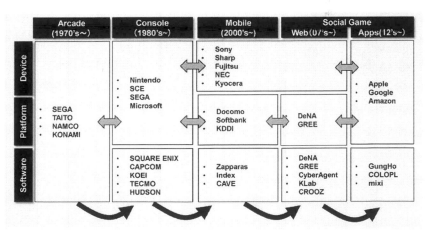

Figure 8. Graph of transitions between eras of platform dominance. Atsuyo Nakayama, *The Third Wave of Japanese Games*, Kindle Location 515.

but nonetheless take the same pattern as Docomo: they operate as platforms for other contents providers, acting as mediators themselves for third-party services, supported by ad revenues and freemium game income. In the process, the contents providers for i-mode become platforms in themselves. They in turn recruit other contents providers and mediate the transactions on their platforms. One platform becomes the generative hub from which other platforms develop and extend outward. This hub formats the other hubs or platforms along similar principles to the i-mode platform; the logic of platform-as-mediator pioneered in i-mode is incorporated into the platforms that grow out of i-mode—including ones that migrate from the mobile to the PC web. I-mode hence can be seen as a kind of genesis moment for the Japanese commercial web in toto. It makes the monetization of contents possible, and serves as a model for later platforms. It also explains in part the discursive effervescence of the contents concept analyzed in chapter 1. For a long while, contents providers could rely on cell phone companies to be the platforms or intermediaries; when these began their decline, the contents providers themselves built the next generation platforms. One of the most important ones in this regard—alongside DeNA and GREE—is Dwango's Niconico Video, to which I now turn.

From Contents Provider to Contents Platform

In the introduction to this book I briefly noted the October 1, 2014, merger of Kadokawa, one of Japan's largest publishers and aspiring media companies, and Dwango, owner and founder of Niconico Video (among other ventures), Japan's largest and most culturally significant video platform. Contents and platforms were the keywords of the justification of this merger, which is why this moment is of particular symbolic significance for this book, displaying the discursive, medial, and financial effects of the platform economy. Dwango founder and former Kadokawa chairman (and now chief technology officer) Kawakami Nobuo explained the reasons for the merger: "Kadokawa and Dwango have both developed contents and platforms—Kadokawa for the real world, and Dwango for the world of the Net. Having offered both contents and platforms for the net world and the real world, the compatibility of these two companies united as one is extremely good."[38] In a statement that hints at the nationalist drive behind the merger, the ways the Cool Japan discourse plays into the sense of Japan-origin contents and platforms as globally valuable, and the fear of American platforms in the Japanese media space, Kadokawa Tsuguhiko enthusiastically described the merger as resulting in the creation of "the Platform of the Rising Sun." The rising sun is of course a nationalist code word for Japan used most frequently during the wartime, imperial era of the 1930s to 1940s, during Japan's imperialist incursion into Asia.

The merger is compelling for other reasons, in addition to the ways it voices anxieties about American platforms. These include not least of all the incredible push by book sellers and publishers to have a foot in the new media world, and the potential overlap in core audiences of both of these companies, namely so-called *otaku* ("geek") male and female fan communities to which both companies cater. It is also useful for thinking what this merger actually is of, in light of the discursive, financial, and social platform economic concerns of this book. At first glance, this would appear to be a merger of a platform company (Dwango) with a contents company (Kadokawa). Dwango is a tech company and owner of Niconico Video, a video platform that encourages the uploading of third-party, user-produced contents. Conversely, Kadokawa would seem to be a contents company, publishing magazines and books, and also having a long history of film and animation production and distribution. Scratch the surface, though, and Kadokawa also appears to be a platform company, perhaps even more than a contents company, not the least because of its investment in magazines (recursively defined as a type of platform), along with the e-reader BookWalker.

Ōtsuka Eiji, a writer and intellectual both closely associated with Kadokawa and vocally critical of the merger, makes the case for understanding this business deal as a merger of two "infrastructures" or platforms. In a critical think piece published a mere three days after the merger was announced, he writes:

> The merger of KADOKAWA and Dwango is not one of contents and infrastructure [i.e., platforms], it is a merger of infrastructure and infrastructure. If it were a *merger* of [Dwango subsidiary] Niconico and [animation studio] Ghibli, or KADOKAWA and [publisher] Kodansha—of

course we shouldn't be surprised if any of these were to merge either—that would indeed be the merger of infrastructure and contents. But this is not the case here. The merger of KADOKAWA and Dwango is not contents and infrastructure, it is the merger of infrastructure and infrastructure.[39]

Moreover, what Ōtsuka suggests is that Kadokawa's infrastructure—what we would call platform—is the *media mix* itself: "What's important here is that KADOKAWA's essence lies in its particular *infrastructure* known as *the media mix system optimized for a limited otaku market.*"[40] This is an interesting shift, one that situates a transmedia practice as a form of infrastructure, or basis for other contents.

Ōtsuka has a point: both companies are associated with platforms, but both are also contents producers. This is indeed what Kawakami Nobuo suggested in an interview: that Dwango is a contents company, not a platform (despite current rhetoric to the contrary).[41] Given its roots as a contents provider for i-mode, this positioning of Dwango as a contents company makes much sense. In his book on the future of the internet and the role of platforms within it, Kawakami also takes the position that platforms should always produce their own contents too—otherwise they risk simply being exploitative rent-seekers.[42] Taken from the present perspective, however, Dwango is also very much a platform company, housing the contents platform Niconico Video and serving as the conduit for other companies' contents in their book reader or official "transaction video on demand" (TVOD) or paid-for video streaming channels.

Indeed, it is the ambiguous position of Dwango as both contents provider *and* later platform that makes it so very interesting. The argument I will advance in what follows is that this dual investment in platform and contents informs not only business practices but also the very aesthetic form its Niconico Video streaming platform takes. For this reason we should view Niconico Video as a hybrid form of the contents platform, wherein the particular affordances of the platform lie in the purposeful disintegration of the firm boundaries between platforms and contents that most platforms—from YouTube to Netflix—firmly maintain. This section thus broaches a topic this book has thus far discussed too little: the aesthetic and cultural lives of platforms, including both the cultural politics of the platforms and the cultural–aesthetic impact platforms have. The politics of platform aesthetics is a particularly pertinent issue in an era marked by the mediation of contents providers by American multinational corporations, which tend to emphasize the transparency of the platform itself.

First, though, let us briefly review the reasons for Dwango's transition from being a contents provider to being a platform service. As we have seen in the first part of this chapter, the increasing power of certain contents providers over the course of the first decade of the twenty-first century allowed some to slowly reshape themselves into cross-platform portal sites—platforms unto themselves. This was as true of Dwango as it was for Mobage Town and GREE. Dwango's pivot to becoming a contents platform came in 2005. By then, the contents Dwango provided had shifted from its early hit games to being entirely based around its Iro-Melomix service, which provided ringtones and later full songs on a subscription basis as of 2001 for i-mode, EZweb, and by early 2002 for J-Sky as well.[43] The service was an instant hit—in part through a marketing scheme that saw it advertised on the high-traffic Maho I-rando (魔法iらんど, or "Magic i-sland") mobile phone novels site. By its first month of activity, its revenue had surpassed that of Dwango's game ventures, and by three months of operation it accounted for 80 percent of all revenues for the company. The success—popular and financial—of Iro-Melomix lasted until around 2005, when the viability of the system was suddenly challenged by a change in licensing agreements with music rights managers that meant Dwango would no longer be able to offer full songs. The result was that starting in March 2005 the membership of the site declined, and Dwango was more or less "defeated" in what Sasaki Toshinao, in his popular account of the company, calls the "full song wars."[44] By that point, 90 percent of Dwango's mobile phone market income came from the Iro-Melomix service, which meant its decline could be a potentially ruinous blow that would have to be met with some fast thinking.[45]

What Dwango ultimately came up with was a plan for a video streaming site with user comments streamed over the video. At first, it aimed at a video streaming site for mobile phones, thinking that the flat-rate packet plans that started in 2003–4 had made this possible. But after some trials

Dwango came to the conclusion that it was too soon for video on mobile phones, and so it pivoted to offering video streaming services for the web.[46] This was the birth of Niconico Video, pioneered in part by the famous internet personage Nishimura Hiroyuki, the founder of 2ch, Japan's most infamous bulletin board site. One part of the Niconico Video origin story is, as narrated above, the shift in power relations between contents provider and telecom platforms. But another aspect of the importance of Niconico as a phenomenon is the manner in which it intersects with transformations in the distribution of cultural goods that attend the rise of contents and platforms discourse, as well as the shifts in the locus and location of platforms—from the Japanese national telecoms to U.S. multinationals like the so-called Gang of Four (Apple, Google, Amazon, and Facebook), more recently reformulated as FANG (Facebook, Amazon, Netflix, Google [now Alphabet]). The addition of Netflix is significant insofar as it is a streaming service that aims toward "global domination of TV streaming services" by operating in almost every country in the world.[47] In other words, Niconico Video is the possibility of an "other" YouTube, a different Netflix, an alternative to the U.S.-dominated streaming paradigm. Characterized by comments streaming over video, Niconico's crowded visual field marks a departure from the increasingly clean, convenient, full-screen contents format on U.S. platforms. While Niconico will likely never become a globally dominant platform, it presents itself as an alternative with a significant user base in Japan, and more important still, a radically different aesthetics of interface. Indeed, as the interface aesthetics have become incorporated into Chinese streaming platforms Bilibili and Tudou, and South Korea's AfreecaTV, the interface aesthetics of Niconico have already had a considerable impact on streaming markets outside of Japan, with implications for a conceptualization of a regional video formation.

The alternative interface aesthetics that we find in Niconico is significant for the ways it syncs up with the Japanese media ecology of the media mix, with which it is more closely integrated than in the case of the U.S.-origin platforms. The media mix refers to the quotidian, ubiquitous, and industrially ingrained habit of linking cultural commodities together into multistrand narrative universes, or, selling books, magazines, moving images, and objects as part of the same media franchise.[48] Within the media mix, one content is transformatively ported to another platform, with each media incarnation used as advertisement for another, potentially fueling a virtuous circle of consumption that props up the entire franchise. The logic of the media mix informs most commercial media production in Japan, and for this reason ties video together with print, games, comics, and novels in a tightly knit ecosystem. This ecosystem operates through particular kinds of mediation, relying on the material specificities of each object (comic frame, animation movement, toy form) as well as the proliferation of a particular franchise through magazine advertisements, material space, and fan production. This materiality is a form of physical mediation that is the motor of the media mix.

As with cultural production in other realms, the materiality of the media mix is undergoing significant transformations, the bywords of this change being *contents* and *platform*. The rise of digital delivery of contents is one site where the cultural politics of platforms negotiate the increasing global platform hegemony of U.S. transnationals. These transnationals such as Amazon, Apple, and Google offer an increasingly homogeneous and unified ecosystem of contents delivered whose video and print interfaces emphasize visual transparency and what I call "full-screen aesthetics," and collectively raise the specter of platform imperialism and an aesthetic uniformity. Niconico offers a counterpoint to this mode of mediation through its emphasis on user interaction via an onscreen, in-video, comments overlay function, developing new connections across contents and platform. In other words, Niconico may offer a more appropriate interface for the textual and material portings and interreferences on which the media mix depends. The tight relation between print and video integral to the media mix (in the form of book or magazine in relation to television or film animation) is reinforced as Niconico delivers not only video but also manga, novels, and magazines, via the same interface as its videos, and in conjunction with its unique comments function.

As such, Niconico presents one possible model of a *counterplatform*. With counterplatform I invoke film theorist Peter Wollen's propositions on a radical countercinema, which he developed in a 1972 analysis of Jean-Luc Godard's political modernism. The fundamental point Wollen makes is that Godard's cinema is political insofar as it highlights the apparatus and ideology of filmmaking, as opposed to the transparency and forced disappearance of the process of production of film within the Hollywood system. This theorization is taken up and transformed by media theorist Alexander Galloway in his subsequent formulation of countergaming (in which he focuses on the modernist tendencies of the internet art duo Jodi. org).[49] While the political valence of their arguments shifts as we examine online video, the argument against transparency in both interventions informs my development of the counterplatform here. Yet as I explain below, my emphasis on the *nontransparency* or *opacity* of the interface differs from their emphasis on foregrounding as the conceptual opposite of transparency. Hence the counterplatform does not operate in the modernist mode of foregrounding the apparatus but rather internalizes and foregrounds the messiness and polyvocality of the web—as well as the media mix—within new distribution platforms. In its opposition to a dominant paradigm of *transparent distribution* privileged by many U.S.-origin streaming platforms, the *transformative distribution* developed by Niconico arguably makes it a model of a counterplatform.

Platform Imperialism

Distribution systems and interfaces, search engines, platform ecosystems, and the legal structures that support all of these are increasingly "made in America," even if their hardware supports are not. Apart from linguistic translation, these platforms operate relatively uniformly at the level of physical and software interface. This is visible on an everyday basis in the increasing global dominance of iPhone and particularly Android devices around the world, monopolizing market share and funneling users toward their proprietary apps and contents ecosystems. This transformation is all the more remarkable in Japan, where its formerly vibrant feature phone culture has given way to the smartphone rectangular slabs of glass. This de facto infiltration of markets by Apple and Google gives the two companies an unprecedented reach into the cultural lives of their users.[50]

The increasingly global domination of the Gang of Four or FANG also suggests that we may want to put the ascendancy of the commercial platform ecosystem into a geopolitical frame, reopening the question of cultural imperialism. Cultural imperialism is a framework deployed in political–economic theories of globalization, wherein critics decried the erasure of cultural differences under the sameness of American culture as it was distributed and consumed around the world (particularly via Hollywood films, American television shows, and fast food culture). Fredric Jameson engages closely with this position, arguing that there "is a fundamental dissymmetry between the United States and every other country in the world, not only third-world countries, but even Japan and those of Western Europe."[51] Through institutions like NAFTA and through GATT trade agreements, U.S. cultural goods are made to circulate through the world, raising the specter of what Jameson calls "the violence of American cultural imperialism and the penetration of Hollywood film and television," which result in the "destruction of those traditions" cultural and otherwise in target countries.[52] This argument is given concrete basis and granular political–economic detail in studies like *Global Hollywood,* in which Toby Miller and his coauthors exhaustively examine the economic and institutional conditions for Hollywood's global cultural domination. By the late 1990s, under the influence of reception studies, cultural studies, and postcolonial theory, critics advocated for a more nuanced approach to Hollywood and American cultural circulation, particularly calling attention to the different manners in which the same goods may be taken up or interpreted. This played up local difference and cracks in U.S. mass cultural hegemony.

Yet political economist Dal Yong Jin reopens the question of cultural imperialism in the context of the increasing economic and cultural platform domination by Silicon Valley giants. *Platform, interface, and hardware* device may well represent a renewed axis of American media imperialism. Jin, who first elaborates the concept of platform imperialism, writes the following.

Accepting platforms as digital media intermediaries, the idea of platform imperialism refers to an asymmetrical relationship of interdependence between the West, primarily the U.S., and many developing powers—of course, including transnational corporations. . . . Characterized in part by unequal technological exchanges and therefore capital flows, the current state of platform development implies a technological domination of U.S.-based companies that have greatly influenced the majority of people and countries.[53]

Whereas Jin focuses his analytical attention at the level of ownership, in what follows I will draw our attention to the aesthetic politics of interface and distribution, and Niconico Video's unique agglomeration of contents and platform. While Japan is by no means a developing power—and operates its own forms of cultural domination in East Asia in particular—it is also a test case for the issue of the sustainability of cultural production in the face of the aesthetic and cultural mediation operated by U.S.-origin platforms.[54] In the introduction to this book I signaled my reservations about the platform imperialism hypothesis, noting Kadokawa Tsuguhiko's nationalist leanings at the Kadokawa–Dwango merger, and the ways that the focus on invading U.S. platforms reinstates the U.S.–Japan dichotomy or cofiguration of the Cold War era, all the while erasing the violence of Japanese corporate and cultural expansionism in Asia. This reinstates what Naoki Sakai has called the "schema of co-figuration" (wherein Japan always defines itself in relation to the West, or the United States).[55] It is here that we see most clearly the pitfalls of the platform imperialism hypothesis; even as it draws attention to the political–economic dominance of U.S.-origin platforms, it also authorizes a cultural and economic nationalism that easily falls into line with the Cool Japan propaganda of the government.

And yet, despite these reservations about deploying the platform imperialism hypothesis in the Japanese context, I would nonetheless like to extend Jin's platform imperialism hypothesis to think through its cultural effects. In so doing I will suggest that platforms and platform imperialism should be understood and analyzed at three levels: (1) interface, (2) contents generation, and (3) distribution. Here I will focus on the level of interface, as it intersects the problem of contents generation in relation to Niconico Video, a platform that plays a major role within the distribution of subcultural media of anime and games, and increasingly operates as a site from which media mixes emerge. Just as the political–economic domination of Hollywood cultural products inspired Godard's countercinema aesthetic (as well as Wollen's theorization of it), so too we must pay attention to the counterplatforms that implicitly or explicitly juxtapose themselves to the dominant Silicon Valley forms. It is in this expanded context that I turn to Niconico, both as part of the wider i-mode effect, as well as part of an incipient counterplatform in the face of increasing U.S.-centric platform mediation.

Niconico Video and the Convergence of Contents and Platform

Founded in 2007, Dwango's Niconico Video is something like the YouTube of Japan; although, according to the internet quantification site Alexa, in terms of sheer numbers, as of August 2018 YouTube ranks as the third most visited website in Japan, while Niconico ranks as the tenth (it nonetheless has the highest daily page views per visitor, as well as the longest daily time on the top twenty sites in Japan, demonstrating its stickiness).[56] Niconico claims about 9 million visitors per month and 3 million daily active users, and it has a premium subscription base of 2.4 million members.[57] This premium subscription base makes it unique among social media platforms—which normally do not have paid memberships—as much of its income comes from this premium base. The willingness to pay for better quality video and other perks also must be seen in the context of the Japanese mobile internet as outlined in the previous chapter. I-mode and other mobile services implanted the habit of paying for services, and hence created a particular model of the platform that is quite unlike most services prior to Netflix—one that assumes users will pay for the operation of the site. With Niconico, monetization happens through subscription rather than user data gathering or advertising, which both have lower priority for Dwango. More interesting still, Kawakami Nobuo explains that not only did i-mode's subscription model influence that of Niconico Video, but its earliest subscriptions went through i-mode.[58] Niconico used the payment method in i-mode as a way of gathering the payments of subscribers, who might otherwise have been put off by entering their credit card information into a new website. Hence Niconico was born out of the i-mode effect, benefiting from its normalization of the habit of paying for services, even as in financial model and payment method Niconico continued to be deeply intertwined with i-mode.

Indeed, Niconico occupies a unique place in the Japanese internet landscape, as it merges the hobbyist internet (generally figured as male techno-enthusiasts) with the mobile web experience, one that is figured so differently that it can be left out of histories of the internet almost entirely. Niconico is a hybrid, for some, combining the "semi-closed services" model of the mobile internet of Japan's feature phones and smartphones subsequently with the cultural cache of the more "underground" internet associated with 2ch.net, which I will discuss below.[59] As such it encapsulates the effects of the mobile web on the web in toto.

After transitioning from its initial incarnation as a site for uploading unlicensed anime and film videos, it became known as a hub for three particular kinds of user-generated contents—virtual idol Hatsune Miku songs and videos made using vocaloid software, THE IDOLM@STER videos made using the game of the same name, and videos using the fan-produced Toho Project—and is known for its close ties to Japan's anime-manga-game subcultures.[60] Perhaps the most emblematic contents in this regard is Hatsune Miku, a vocaloid software program produced by Crypton Future Media that allows creators to use the voice of a character to make their own music. What is unique about this program is that the initial software box came adorned with a particular character design for Hatsune Miku, a lithe girl in a short skirt framed by two long pigtails of green hair. Creators used the voice to produce music, but visual artists also drew on the iconography of the Miku character to create accompanying music videos. Miku quickly became the object of fan creation, both of music and of visual animation to match the music. As Ian Condry argues, "The phenomenon around Miku shows that the character, more than the music software, is the platform on which people are building."[61] Miku became platform for the creation of music and visual designs, and Niconico Video was the platform of choice to upload, share, comment, and modify Hatsune Miku creations. In recent years, Niconico has diversified from the Miku videos, launching subsections for the website, featuring video-on-demand (both by subscription and one-off purchase) and an interface to read novels and comics (both professional and amateur) called Niconico Still Images (Niconico Seiga). It has also launched a Niconico live section where it has broadcast exclusive speeches by politicians during election campaigns, and it also hosts less official fan-made live broadcasts.

The most defining feature of Niconico video is without question its video-overlay comments function. Unlike YouTube, Niconico is not only a site where you upload and view user-generated contents (UGC). It features an interface that operates as a means of transforming these contents, namely through the video-overlay comments function.[62] As such, Niconico is an entirely unique video streaming site: it is a platform that also creates contents.[63] It allows for contents to be uploaded and shared, but in the process of commenting, new contents are created. Dramatic sections of videos are adorned by stars, emoted by series of *w*'s designating laughter, and hermeneutic interpretation unfolds in the comments in-video. Comments write affective tonality and interpretive gestures into the fabric of the video itself. In brief, the comments are the source of innovation, media connection, and community formation, and the origin of a language shared by users about media that in turn informs media.

Sasaki Toshinao in his popular account of Dwango describes the comments as follows: "In other words, the comments flowing on screen are not simply 'add-ons' or freebees. The comments and video merge to generate new contents. This is a new world where moving image contents and social media have converged."[64] Niconico develops the convergence of contents and platforms, where the platform's affordances—the comment function—allow for the transformation of the moving image itself. This stands in stark contrast to platforms like YouTube where the uploaded videos are by comparison static, unchanged by user interaction, with the platform operating as a transparent interface to the video content. Comments are placed below the video and are merely supplemental. It is in this sense that Niconico operates as both contents producer and as a platform. Niconico is not just a platform-as-delivery mechanism of content in the Amazon-Google-Apple-Netflix vein.[65] Niconico is a space of interactive creation where contents are transformed by their process of distribution. Distribution becomes the site of active cocreation, a node where contents and platforms converge. Of course, this could be the object for a critique of fan labor—something I will return to later in this chapter.[66] For now, however, I would like to offer a redemptive reading of Niconico as a site of potential aesthetic resistance to platform imperialism and its transparency

of content delivery and display, occurring at the surface level of interface—
or *intraface*.

In his useful discussion of interfaces, Alexander Galloway proposes the
term *intraface*, which he defines as "*an interface internal to the interface.*"[67]
The intraface offers Galloway something of an intermediary term between
transparency and foregrounding found in debates around political modernist cinema—a binary between erasure of mediation and its disintegration
that he usefully seeks to complicate. As an example of the contemporary
intraface, Galloway points to the image-text juxtaposition of the online
game World of Warcraft, noting: "It is no longer a question of a 'window'
interface between this side of the screen and that side . . . but an intraface
between the heads-up-display, the text and icons in the foreground, and
the 3D, volumetric, diegetic space of the game itself—on the one side, writing; on the other, image."[68] This internal interface, Galloway continues,
indicates the presence of "*the social*" within the frame.[69] Adopting Galloway's useful formulation of the intraface—which describes perfectly the
text-video (or writing-image) environment of Niconico—I would like to
nonetheless shift from an analysis of "*the*" social (transnational capital) to
"*a*" social: the very particular media ecology around Niconico that points
on the one hand to Japan's net culture and on the other to the connectivity of the media mix. Rather than move from *the* social to an allegorical
reading of a text (WoW), I prefer instead to move from *a* social to its
medial and historical context—deuniversalizing the web space in a manner that echoes critiques of the rhetoric of the idealized, global internet
and calls for analysis of national or regional particularities. This approach
acknowledges what Michelle Cho calls "the existence of multiple internets
that are determined by language, regulatory measures, software interfaces,
and communications infrastructures" resulting in a web that operates as
"an uneven engine of mass identity formation."[70] In addressing Niconico,
we must turn our attention to the comments, the communities that emerge
around them, and the way these become part of the tissue of the media
mix. This requires us to focus on the particular form of opacity that Niconico both relies on and further develops.

Here I use *opacity* as a term meant to contrast the emphasis on transparency and full-screen aesthetics prevalent in video platforms such as

Google Play, YouTube, Amazon Prime Video, and Netflix, all of which default to full-screen, comments-free video display. In these the cinematic full-screen video mode takes precedence over the televisually inflected multitexual environment we find with Niconico. Shunsuke Nozawa first develops the concept of opacity in relation to the particular forms of anonymity specific to Japanese online bulletin boards, whose culture Niconico inherits. He suggests that "crucial to the structure and experience of Japanese virtual communication are *acts of opacity:* the presentation of the self-in-disguise."[71] Daniel Johnson further develops the concept of opacity to refer in particular to the illegibility of text in Niconico videos—what he calls the "non-denotational, counter-transparent typography of deliberate mistype and production of sprite-like images in Nicovideo commenting."[72] To these two types of opacity—anonymity, and illegibility of the comments—I would like to add a third type: the opacity of the image, and the illegibility of the distributed contents. Moving from the political modernist dualism of transparency/foregrounding (hiding versus showing the apparatus of production in film) to nontransparency and opacity allows us to sidestep the either/or logic of ideology critique and unveiling that is embedded in these terms, and which is ultimately of limited use for thinking the particular consequences of the opacity of the decidedly commercial and only ambiguously oppositional Niconico. After all, Niconico's opacity is not about deconstructing commercial media and the media mix but rather about doing the media mix differently. It is therefore in line with the particular aesthetics and messy politics of Japan's net culture.

Situating the particular form of nontransparency of Niconico's intraface requires an acknowledgment of the impact of Japan's infamous 2 Channeru (2ch.net)—known as 2chan *(nichan)*—on the visual interface and community of Niconico. For it was the commentary culture of 2ch and the visuality and authorial opacity (or anonymity) that influenced the interface design of Niconico. This is a meeting of two very different internet cultures: 2chan's free-for-all "wild west" of Web 1.0 and the paid services ethos of the i-mode internet—both of which emerge in the same year, 1999, and yet have radically different trajectories, rejoining in the form of Niconico Video.[73] Yet a discussion of the form of 2ch cannot be dissociated from a discussion of its politics, and in particular those of the net-right (*netto uyoku*) with which it is so closely associated.

A simple bulletin board site launched in 1999 by Nishimura Hiroyuki, 2ch quickly became one of the most significant and sometimes scandalous sites in the Japanese net. It was and continues to be anonymous, and completely text-based, with threads limited to one thousand posts. Its imageboard imitator Futaba Channeru (2chan.net) is in turn the model for the infamous English-language site 4chan, from whose anonymous threads the hacker collective Anonymous takes its name, where gamergate first unfolded, and in which the fascistic alt-right found its feeding grounds. Its politics are oppositional but in a manner that illustrates the dangers of populism and its profound failures in relation to progressive politics. Like the boards of 4chan, which anthropologist Gabriella Coleman describes as being "unique for its culture of extreme permissibility," 2ch.net developed its own language (or linguistic *détournements*) and aroused much controversy.[74] In particular, 2ch.net became known as a hub for the net-right or right-wing expression on the internet—much like 4chan would become with the alt-right.[75] It is impossible to ignore the disconcerting rightward swing of Japanese politics over the last decade—and 2ch certainly has some responsibility for this shift.

That said, some analyses of the net-right phenomenon have sought to lend complexity to the sometimes knee-jerk condemnations by the mainstream press and the left. One of the most compelling of these is undertaken by media theorist Akihiro Kitada, who describes the type of communicative behavior on 2ch in the following manner: "The most typical communication style on 2ch is trading snarky commentary on specific kinds of source material. The communication is intimate but harsh; the harshness is itself a kind of intimacy."[76] Kitada continues: "From a sociological point of view, 2ch marks a departure from an instrumental rationality that supports the existing social order. Instead, it is a social space that produces an extreme form of *connective* rationality that supports ongoing communicative actions and reactions that maintain the community."[77] Kitada proposes that the net-right's communicative utterances be seen at least in part as a cynical reaction to the mainstream mass media (whose literal-minded condemnations of 2ch provide fuel for their fire), which involves simply saying what should not be said, and in part as a function of the first order of communication according to Niklas Luhmann: keeping communication going (under which principle net-right speech functions as an effective means of provoking further communication, much in the same way trolls flame wars and enjoy the communicative burn).[78] Kitada hence historicizes the emergence of the kind of expression in which communication is more important than content—"where media content is a servant to the maintenance of this particular form of communication."[79] It grows increasingly difficult to maintain the critical distance Kitada adopts in his early twenty-first-century intervention, as the metaprovocations and counterpositioning of 2ch and its denizens settle into habituated right-wing positions. These then feed into the culture that grows up around Niconico Video, even as the latter in recent years has attempted to clean up its image.

I linger on 2ch in part because its founder and iconic figure, Nishimura, was a board member of the parent company of Niconico, Nwango (a subsidiary of Dwango), and very present in the development of the video service. Nishimura was also partly responsible for the initial popularity of the site, as 2ch denizens flowed over to Niconico video upon his recommendation. If Nishimura's recommendation initially pushed 2chanellers to the site, the stickiness of Niconico can be explained by its similarities to the quasi-live feel of 2ch. The creators of Niconico—a small group of programmers, along with Kawakami and Nishimura—took the "live coverage" *(jikkyō)* bulletin board on 2ch as a model. The live coverage board is a place where users congregate and exchange live commentary on particular television programs airing at that time. The programmers for Niconico took this as a model, and decided that commentary should scroll on top of the image. Following 2ch, Niconico limited the number of individual comments to one thousand.[80] Unlike 2ch, the Niconico comments are not "live" or in real time. Instead, Niconico deploys what Hamano Satoshi and others have dubbed "pseudo-simultaneity,"[81] where the position of a given comment on the timecode of a video remains constant. (For instance, a comment I make at the 3:05 time-code mark will be replayed there for anyone who watches the video subsequently, until it is replaced once the one-thousand-comment limit is reached and new comments cycle in to replace mine.) Comments are seemingly simultaneous in relation to the particular moment of viewing the time-shiftable, replayable video. This allows conversations to develop across time but around a particular time-code mark, and becomes a reason for commentators to revisit videos: to see what other people have said in the meantime, to see if their comment still remains, and perhaps to leave another comment. In keeping with the structure of 2ch, the comments are anonymous, and the site operates through pseudonyms.

Niconico's initial affinity with 2ch and proximity to the 2ch community has endured, even as the user base for Niconico has expanded outward; the platform has particular forms of vocabulary associated with it, idioms that form and are replaced over time, and shared cultural codes when commenting on videos. Unlike 2ch (or 4chan), the tone is not uniformly snarky or cynical. Indeed, users are often very encouraging to video up-loaders, especially inexperienced ones. The limits to hospitality can be found in videos about politics, especially videos sympathetic to or defending resident Koreans in Japan, where the net-right's litany of insults returns in full fury—giving it a reputation for being a racist (and troll-friendly) platform not so far removed from 2ch.

Yet amid this unpredictability, Niconico presents a site of creation, response, and recreation that harnesses on the one hand the unruliness of net culture coming from 2ch and on the other hand the creative and recre-ative forms of cultural production centered around the famous physical site for the exchange of so-called secondary production of manga, anime, and game culture: the Comic Market. To these two, it brings legally streamed official contents—from anime shows, to TV dramas, to political speeches, all of which are fodder for further comments. It offers itself as a prime candidate to examine the potential effects of a site that literally flaunts mediation on its nontransparent, opaque intraface. By nontransparency I refer to the addition of elements to the image rather than an unveiling of elements within the image. The "non" here is a transformative addition, an additive opacity or surplus of meaning rather than a revelatory subtrac-tion. And what is added, of course, is in part the community of commen-tators and the affective tone of comments. It models how a platform can be more than a site of distribution, a quasi-transparent point of access to what—in a variation of Gilles Deleuze's formulation—we might call "any-content-whatever."[82]

Niconico highlights the potential of a platform that caters to and mediates smaller-scale, regional, subcultural *contents* networks whose preservation depends on their filiation with appropriate platforms ecosystems.[83] Here we can see Niconico's affinity for the subcultures of manga, anime, and games whose communities it fosters. Niconico develops an ecosystem of fan-based production that can more porously evolve into new forms of the media mix. The media mix operates through a series of mediations, encounters with contents, as well as transformations of these contents, whether this be novel, film, anime episode, magazine special issue, online game, and so on. Insofar as Niconico, along with its community of contributors and commentators, is structured around the continual engagement and modification of media objects, it presents itself as a platform optimized for the continuation and evolution of the media mix. The more open architecture of Niconico reflects the active model of fandom at the core of the subcultural media of anime, manga, and games in Japan. Moreover, the smaller scale of the community also brings to the site the sense of common participation in the distribution, mediation, and interpretation of cultural forms called contents. Niconico as a platform operates as a form of marked mediation (through on-video comments) that encourages further creation. And this further creation is part of a cycle of active consumption that Niconico harnesses. A crystallization of this function in Niconico as media mix generator can be found in the Kagerou Project, a media mix that emerged from Niconico and peaked around the time of the Kadokawa–Dwango merger.

Kagerou Project: Platform-Based Media Mix

The Kagerou Project initially debuted as a series of songs uploaded to Niconico. The songs were created by the previously unknown producer Jin, who used vocaloid software (sometimes using Hatsune Miku, sometimes other software) to make a cycle of songs in which each tells part of a larger narrative. Jin wrote the songs, and his visual collaborators Shidu and later Wan'nyan-pū created the visuals and character design in a series of informal collaborations that was central to the development of the Kagerou Project (see Figure 9). The songs themselves act as micronarratives, and

Jin wrote a series of light novels illustrated by Shidu using the songs as the basis for an expanded narrative. This expanded narrative became the basis for both the manga and the anime series, all of which draw on the world created in the songs and novels.[84] This use of a song cycle to unfold a larger narrative world was unique, and the increasingly popular songs quickly evolved into a formal media mix that included CDs, light novels, manga, magazine serializations, and an anime series, all of which Jin was closely involved in as author or producer. The Kagerou Project itself became the basis for official and unofficial secondary fan works produced around it, a selection of which were published by Kadokawa imprints Media Factory and Enterbrain. Jin created the songs but also wrote the novels (published by Enterbrain) and is credited as the creator of the original work that is the basis for the comics and the anime. The anime series—titled *Mekakucity Actors* (2014)—was produced by the noted animation production unit Shaft and was overseen by anime auteur Shinbō Akiyuki. Known collectively as the Kagerou Project, this is one example of a media mix that emerges out of Niconico, develops across media, and where most of the elements or products of the mix are accessible from within the Niconico platform.

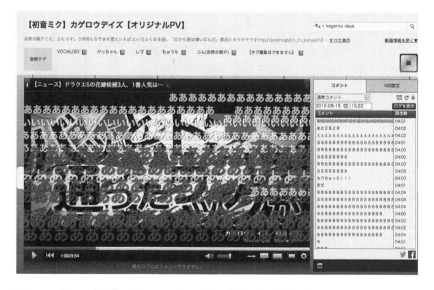

Figure 9. Screenshot from Kagerou Daze video. Niconico Video, January 7, 2015.

In short, what we see here is a new model of media mix that grows out of user-generated content, harnessed and further developed by Kadokawa and Dwango. This is the prototypical form of a media mix that emerges from out of the informal networks fostered by the Niconico site and expands into a formalized media mix. Moreover, almost all aspects of this media mix are then available for viewing, reading, or listening on Niconico. We can watch the original videos, pay to watch the anime, and buy and read the comics and novels. And here we begin to see the endgame of Niconico as a platform: it has moved from a site that predominantly hosts videos (including paid streaming channels) to a site that also hosts its own manga and novel bookstores and e-readers. In line with the comment culture of the video, Niconico Seiga (or "still image") adopts the same interface as Niconico Video. Comments stream across the manga or novel page as one reads, allowing viewers and viewing-commenting habits, dispositions, interpretive dialogues, and emotive responses to migrate from one part of the site to another (see Figure 10). In keeping with the main feature of the site—the over-the-image comment function—books and manga can be read both for the existing contents and the social content of

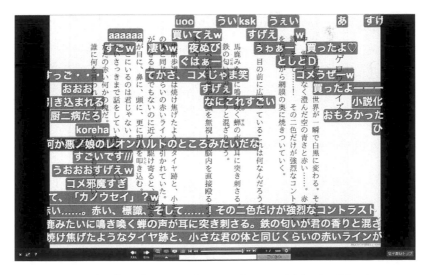

Figure 10. Screenshot from first pages of Kagerou Daze novel. Niconico Seiga, January 29, 2015.

the comments. This is social reading, mediated by an interface that encourages a smooth move from one aspect of the site to another, from free video, to paid streaming, to paid reading.

Jin and the Kagerou Project hence offer something of a roadmap to how Kadokawa and Dwango in their conglomerated form seek to synergetically operate to harvest user-generated content and unfold media mixes across their various properties, from video to books to manga to anime. This alliance suggests that the nontransparency of the interface that Niconico as platform provides is congruent with the tissue of relations that make up the media mix ecology. If media mixes work in part through the mediating and connecting function of the character, Niconico raises the prospect of a media mix working through the connecting function of the comments themselves. Countertransparency here works in the service of commercial deployment and draws attention to the ambiguous status of user-generated contents as incipient media mix material. At a textual level the Kagerou Project emerges from the intraface of Niconico and reminds us of the generativity of the interface. It shows too how Niconico may be a counter-platform in relation to the U.S.-origin platform models in its aesthetics of interface, but it is no less a commercially driven platform for this.

From Toyotism to Dwangoism, Or, What Is Platform Capitalism?

Niconico's particular kind of mediation between contents and platform registers its earlier history as a contents provider for i-mode. The emphasis on the generative potentials of the interface and the intraface as sites of contents creation and distribution reflects a continuing commitment to contents creation by a company still very much aligned with its original position as the creator of contents for i-mode. The personality of the contents provider is baked into Dwango's post-i-mode-effect transformation to platform. In its early days, Niconico's premium-based subscription model was, as we saw, supported by the i-mode payment system. There is hence a platformic transfer of powers here, a platformization of the contents provider, albeit one that comes in an era of the increasing dominance of American platforms in the mediation of life. Formal and informal alliances are developed with other cultural platforms. These alliances include formal ties to Kadokawa's ecosystem of media mix contents and its ebook reader such as BookWalker first and foremost but also informal ties to streaming sites in China and Korea that have also adopted a similar comments-over-video format. As Jinying Li notes, China's anime streaming sites Bilibili and AcFan first introduced the comments-over-video feature, and its giant streaming service Tudou quickly followed suit. As Li writes, "By 2014, almost all major video-streaming platforms in China feature a *danmaku* [comments-over-video] interface, which is no longer a unique subcultural entity but has become a standard interface design in Chinese online video culture at large."[85] Chat app WeChat and microblogging platform Weibo have incorporated over-video comments into their social media services. Even movie theaters in China have experimented with contents over video.[86] In Korea, the AfreecaTV streaming platform allows bubble emoji and other image overlay over video (which, uniquely, is also part of its monetization structure), a feature that Facebook has since integrated into its own video streaming.[87] Niconico Video has thus had significant regional effects on the structure and interface of streaming services, even if it itself remains a predominantly Japanese platform in terms of its own user base.

The interfacial specificity and generativity of Niconico has led its enthusiasts to posit it as being a new industrial model for Japan moving forward. In a passage that resonates with the industrial and economic platform lineages traced over the course of this book, Hamano Satoshi, the most perceptive theorist of Niconico, speaks of his hope for Hatsune Miku–style generativity and Niconico to become something like a *production model* for the future. In a talk he gave at a live convention for Niconico enthusiasts, Hamano said:

> In the past Americans have keenly studied the method behind the manufacture of Toyota cars, under the various appellations of "the Kanban method," "just-in-time," or so-called "kaizen" [improvement]. I'm sorry if I'm going out on a limb here in this comparison to Toyota, but what I would like to see moving forward is for Japan to become known for what I will call the "Niconico/Miku style" manufacturing method. This is a manufacturing method that Dwango's Niconico Video and Crypton Media's Hatsune Miku bring about by heightening users' creative capacities and allowing for the creation of contents and software.[88]

On first read these comments may seem farfetched, as Hamano himself seems to recognize. Dwango and Hatsune Miku could replace Toyota—really? And yet this comparison is also fascinating, recalling other such gestures we have seen over the course of this book. It replicates the narrative of Japan's necessary shift from automobile production and the manufacture sector to cultural creation and the creative industries, a hope for an economic savior embodied in the term *contents*. It also recalls Natsuno's celebration of platforms-as-mechanisms, and his admonishment that Japanese industry should pivot from a logic of "making things" *(monozukuri)*—its famous proficiency for which industry discourse celebrates the success of Sony, Toyota, and other Japanese corporations—to "making mechanisms" *(shikakezukuri),* in the vein of creating entire ecosystems of engagement, in the manner of i-mode and Apple's App Store.[89] In short, these comments see the potential for the creation of culture and schemes adequate to the platform economy. Hamano and Natsuno's respective comments can of course be critiqued for their ties to government policies such as Cool Japan, and for the impossibility of the contents industry, large though it may be, ever replacing the critical role manufacturing had for the Japanese postwar economy. And Hamano's gesture toward Dwangoism is fascinating for its recapitulation of Natsuno Takeshi's own hopes and dreams for i-mode's promise in the early twenty-first century, when he writes in his 2000 book: "When I look at the mobile phone industry now, I feel I am seeing a replay of what happened when Toyota and other Japanese auto manufacturers made their move into the American market in the 1980s. . . . The Japanese mobile phone industry is, I think, staring at an opportunity to conquer the world."[90] An i-modism to match a Toyotism, perhaps?

This Japanese takeover of the mobile world did not pan out as planned. (It did, however, happen in a different form, under the duopoly of Google's and Apple's smartphone systems.) Yet the reference to the auto industry in both Natsuno and Hamano's remarks is fascinating for another reason: they remind us of the central place of the automobile industry within the product-technology platform literature discussed in chapter 2. There I noted that the term *platform* had a first life within an analysis of the automobile industries during the 1970s, 1980s, and 1990s, roughly synchronous with the many analyses (predominantly from the United States but also from Europe and eventually from Japan) of Japan's seemingly indomitable auto sector and the transformations in manufacturing processes that were to embody the shift in capitalist mode of production from mass manufacture to batch manufacture, from Fordism to post-Fordism or "Toyotism" (as it was called), and from an age of uniformity of products and Fordist production methods to an era of diversity in product design and a flexible, Toyotist mode of production.

The platform may have played a small role in articulating this transformation in managerial literature around the automobile industry, but it was a key component of the redesign of manufacture processes around the standardization of parts of the automobile to allow for diversity at the level of outer design. It is also now indisputably a key term for rethinking the economy at large in the era of platform capitalism. In short, the platform expands from the automobile sector and its pioneering transformations in its mode of production within product-technology platforms, to the contents platform, and the platform as transactional intermediary that has become the dominant type of the current moment. Natsuno and Hamano's "from Toyota to i-mode" and "from Toyota to Dwango/Niconico" narratives are surely symptomatic of a desire for Japan's recognition on the world stage. But they are more interesting for the ways they each—inadvertently, perhaps—point to the longer history of Japan's crucial role within the development of platform theory within managerial literature, and the crucial role this theory had in shaping Japanese and global iterations of the platform economy that this book has charted.

Hamano invites us to consider how Dwango's Niconico Video and Crypton Media's Hatsune Miku—both platforms in their own way, as Ian Condry has argued—are now the infrastructure for a different kind of creative activity, as well as the model for a new mode of manufacture, much like Toyota was in its day. Here the contrasts between the two are instructive. There is a shift from paid lifetime employment under Toyotism (which, for all its discourse of just-in-time flexibility, operated half within a regime of guaranteed employment and half with a precariat of subcontractors) to unpaid and precarious labor for the platform laborers of Dwango's Niconico video. This latter precarity reflects the perils of platform capitalism, particularly as publishers like Kadokawa adopt Dwango's model of user-produced products (launching and relying on websites for user-made novels such as Kakuyomu for user-sourced books and crowd-sourced popularity), thereby rendering the creative process even more precarious than it was already and more dependent on unpaid and uncompensated activities.[91] In this regard, Kagerou Project is proof that Niconico is media mix friendly but also that creative labor can be outsourced to user-produced contents.

This making-precarious of previously remunerated creative labor reminds us yet again how the discursive and financial aspects of the platform economy are intimately bound up with transformations in capitalism. Dwango's Niconico presents itself as a counterplatform at the level of interface, offering a critical vantage point for engaging with the encroachments of Google, Apple, Netflix, and Facebook as increasingly global and dominant platforms. And yet, at the same time, Niconico also contributes to an economics of precarity and user-dependent production (dependency is arguably a more accurate framing of this system than the more oft-used term *generated*), within which regime these contents platforms are more alike than they may at first seem. Niconico Video is a unique player insofar as it emerges out of the i-mode contents bubble, and it is one of the contents providers that pivots to being a platform in its own right. This pivot is worn on its sleeve, in its very interface, and in its greater care for the generativity of both users and contents. Yet its dependence on forms of immaterial labor places it firmly within the precarious regimes of un- or underemployment that characterize the contents industry among others within platform capitalism. That its CEO Kawakami unabashedly treats the armies of NEETs (Not in Education, Employment, or Training) as his free day-laborers or reserve armies within his account of the net natives in his book on the future of the internet is symptomatic of Niconico's dependence on unpaid user production.[92]

Hence, we must conclude that while interface design fundamentally differs between Niconico and YouTube—a difference that has true cultural and aesthetic implications—nonetheless the position of the platform as a mediator and intermediary within financial transactions means that Niconico is simply a different model of platform capitalism—one that follows within the i-mode effect—rather than a resistance to it. A countervailing tendency to platform imperialism in cultural and aesthetic form, Niconico is firmly part of platform capitalism within its economic form and its managerial logic. In fact, it is itself an object of nationalistic posturing, as we saw Kadokawa Tsuguhiko doing around the time of the merger. At moments like this, the platform imperialism hypothesis seems to offer cover for overt nationalist jingoism. In this sense, Niconico and other Japanese platforms are as much a global problem as they are regional articulation.

In this sense Niconico operates within the regionalist paradigm elaborated by critical area studies scholar Leo Ching, who argues that regionalism is essential to global capitalism rather than resistant to it. The region for Ching—as I detail in the next chapter—is ultimately not a fixed cultural or geographical entity but a geopolitical entity that mediates the contradictions between "the immanent logic of capital and the historical formation of nationalized economies."[93] Ching writes that although "regionalism may at times appear to oppose globalism, the regionalist imaginary is fundamentally complicit with the globalist project."[94] In its appeal to the cultural specificity of the Japanese media sphere, Niconico is, on the one hand, sympathetic as a local media ecology that runs counter to an aesthetics of transparency seemingly inherent to an imperial American platform mode and, on the other hand, problematic as an emblem of a local Japaneseness that also becomes an emblem of cultural nationalism. Culturally specific and important insofar as it remains part of a specific history and aesthetics of the Japanese internet, demonstrates a support platform for a media mix system of textual ties between elements of the mix, and is a reminder of the continuing effects of i-mode's function as a hub for contents producers that become platforms themselves, Niconico is nonetheless still very much a part of a shift to platform capitalism, albeit one with regional variations.

In its continuation of a logic of platform capitalism present as industrial platform within Toyotism (and despite real shifts in the status of employment noted above), Dwangoism as regional capitalism returns us to the continuities between Toyotism and Dwangoism rather than a break. Not so much an epochal shift *from* Toyotism to Dwangoism, Niconico is a hinge that shows the platform continuity *of* Toyotism and Dwangoism. True, one refers to the layered model of the industrial platform of the car and the other to the contents platform and mediation platform of streaming video.

Yet insofar as it represents an exportable, iterable mode of creative contents production destined to move around the East Asian region, marking itself as a Japanese cultural–industrial export, Dwangoism as a post-i-mode effect platform in the 2010s complements and extends the flexible Toyotist logic that was the industrial–economic emblem of 1980s Japan before it.

Finally, these continuities allow us to finally confront the status of the term *platform capitalism* as deployed in this book, and as used by Nick Srnicek and others. Writings on platform capitalism thus far have been ambiguous as to the precise relationship between Toyotism or post-Fordism and platform capitalism; does the latter signal a new era and thus a subsequent phase of capitalism? Or is it instead a variant on the existing post-Fordist model of capitalism? Does platform capitalism replace post-Fordist capitalism?

My response is that ultimately we should see platform capitalism as a variant rather than a replacement of post-Fordism, building on the continuities we find between Toyotism, i-modism, and Dwangoism. Much as Anna Tsing has theorized of "supply chain capitalism" as part of post-Fordism, platform capitalism should be viewed as an extension or specification of a particular aspect of post-Fordist capitalism.[95] It points to something specific about contemporary post-Fordist capitalism and its regimes of accumulation. Namely, platform capitalism designates what I earlier called the *expansion of managerial power through the principle of market mediation* that results in the reorganization of the social field. But it does invite us to see it as a split from post-Fordism. By qualifying capitalism as "supply chain" or "platform" we are able to point to particular, sometimes new features that extend a logic of flexibility and just-in-time and precarity that were already part of articulations of post-Fordism, this time refigured under the total managerial logic of the mediatory platform. As we saw in chapter 3, platform capitalism (or Dwangoism and i-modism as variants of this) requires a shift from looking at *a* particular platform object (consoles, social media sites, etc.) to seeing the expansion of *platform relations* themselves: the expansion of the management of mediation in the foundation of the platform economy. From an earlier focus on the relations and organization within a given company, attention shifts toward the management of relations between companies, and between companies and consumers, such that all "sides" of the platform's multisided market come under the control of the dominant platform company (which tends toward monopoly, or monopsony). As such, there is a social permeation by a platform logic, one that merits the use of the qualifier *platform* to describe contemporary capitalism. Toyotism, i-modism, and Dwangoism (proper-nouns-made-periodizing-terms) merely point to the crucial roles that firms themselves may have in the expansion of a managerial logic from one organization to the social whole.

In sum, platform capitalism (like Toyotism or Dwangoism or i-modism) suggests an emphasis on a particular *aspect* of post-Fordist capitalism—the total mediation of life by platforms, via intimate devices such as mobile phones and the ecosystems they bind us into—without suggesting the need to declare it a new era of capitalism. Just as we may prefer to speak of *capitalisms* or *neoliberalisms,* acknowledging the local, national, regional variations on global capitalism or neoliberalism, without suggesting that these are entirely different paradigms of the terms, the differences and continuities within platforms suggest that while Toyotism and Dwangoism each bring out different aspects of post-Fordism, we have not entered an entirely *new* era of capitalism. This of course does nothing to obviate the urgent need to understand what these platform-induced qualitative shifts to contemporary capitalism are, where they have come from, and how they will continue to shape our lives moving forward.

The Platformization of Regional Chat Apps

This book has offered a Japanese account of the genesis of the platform economy in concept and practice that in its characteristics, influences, and development thereafter opens onto and impacts the current qualities of platform capitalism on the whole. More specifically, it has offered an account of the definitions, histories, politics, and principles of platforms and the platform economy from the vantage point of early Japanese platform theory and practice as they impact the development of the mobile internet first and the wider commercial internet subsequently. It follows the unfolding of this platform practice and its impact on the conjoined domain of contents, for which I have also endeavored to offer an account—both of contents discourse in its own historical development, as well as contents discourse as it has subsequently been propped up financially and discursively by the rise of platforms like i-mode. We have also followed the intertwined story of platforms and contents through to subsequent developments in the mobile internet, streaming services, and, in this chapter, chat app development.

In this endeavor to offer a Japanese account of the platform I have covered much ground. I briefly recap the trajectory here. *The Platform Economy* starts with the premise that words affect worlds. Lazzarato fuses *monde* (world) and *mode* (mode) in writing that "capitalism is not a mode of production but a production of modes, a production of worlds *[une production de mo(n)des]*."[1] He thereby relates capitalist modes of production

to the production of worlds. In this book I modified his formulation, arguing that *capitalism and language cooperate in the production of wor(l)ds,*
showing how words operate on, model, and cocreate the business ideologies, strategies, and mechanisms that shape our world. The first part of
this book focused on words, the second on their unfolding in practice in
our world. Both are concerned with the ways managerial logics of the
platform in its intermediary function enable the expansion of control over
all of the social field. By being the managerial mediator, the platform operator asserts power over all entities it mediates—including companies, users,
and society itself.

I began by charting the early impact of contents discourse and its permu-
tations, suggesting that contents discourse was an early response to the digital dematerialization of media—and the industry-motivated repackaging
of these into sellable commodities (chapter 1). Second, I gave an overview
of the diverse meanings of the term *platform,* their discursive derivations,
and the sites from which these emerge (chapter 2). Attention here was paid
to the industrial arenas in which platform theory develops, from computing, through automobile manufacture, to software, gaming consoles, and
social media sites. Each of these have for the most part operated within
disciplinary and discursive silos, and the aim here was to both differentiate typologies and draw through lines—particularly tracing the vital cross-
over between computing and automobile industries. Third, I analyzed
one of the earliest Japanese theorizations of the mediation platform that
emerges in the early 1990s, and presages in fascinating ways the type of
platform theories that would emerge in the United States and France in
the early twenty-first century (chapter 3). In so doing, I provided a counternarrative to the predominantly North Atlantic or Anglophone accounts of
platforms, building both on the calls for the internationalization of internet studies and the need to develop a more global vision of media theory—
here figured as management theory. Fourth, I tracked the manner in which
these theories unfolded into platform-building practice in an analysis of
one of the most significant developments of the platform concept in its
time: i-mode and the mobile internet services deployed by its competitors,
au and J-Sky (chapter 4). I-mode in particular was framed by local and
foreign commentators as the future of the mobile phone, the future of the

internet, and the future of mobile commerce—including online purchase of physical goods, the contents market, and the mobile wallet. All of these installed an experience of the mobile internet as a site of paid-for contents and services, habituating users to always-on devices and paid-for contents. This future has materialized in the form of the smartphone, a market now dominated by Apple and Google, with Google's major partners on the Android platform often relegated to second-tier status—namely handset manufacturers Samsung, Oppo, Xiaomi, LG, and so on. While the smart-phone's current incarnation is considerably more axed around the collection of user data in turn mobilized to support other services within the given ecosystem, the contents market, the app store model, and the mediatory operations of the platform all rip a page from the i-mode playbook. Fifth, I mapped the unfolding of the platform in the increased platformization of contents providers (chapter 5) in Japan. There we followed the rise of social gaming sites like GREE and Mobage, which ultimately circumvent the telecom platform, building themselves into hubs unto themselves. Finally, I addressed the cultural function of these platforms—examining Dwango's Niconico Video in particular—in contradistinction to the dominant, U.S.-origin platforms of the day, developing an analysis of the counterplatform in the era of platform imperialism. The chapter concluded with a reflection on the use of Toyotism, i-modism, and Dwangoism as stand-ins for the Japanese economy, and with a reflection on the periodizing term *platform capitalism* itself.

The problem of platform imperialism as conjoined with platform capitalism brings us into another era, and another arena into which we will venture in a preliminary manner in this conclusion. The nation was the clear site and space of the mobile internet platform in the first era I consider in this book—Docomo's i-mode—being bounded by the national orientation of telecommunications infrastructure and its operations. To be sure, Docomo acquired stakes in telecoms in the United States, Europe, and Asia, but on the whole its projects to roll out its internet-enabled phones there did not result in the adoption they had hoped for—leading Annabelle Gawer and Michael Cusumano to somewhat dismissively call Docomo a "platform-leadership wannabee," as if success in the North Atlantic was the sine qua non of being a platform leader.[2] In the end, though, it was Apple and Google's displacement of i-mode feature phones by smartphones that led to the global expansion of the platform model i-mode pioneered (albeit pioneered with a good deal of hints taken from the 1990s wired internet giant, AOL, and Nintendo before that). Under the two American companies, the platform no longer operates in a bounded manner, circumscribed by the nation-state in the way it was during the telecom-dominated first era of platforms. This second era of mobile platforms sees the global reach that Docomo dreamed of—and attempted to accomplish through multiple investments in American, European, and Asian telecoms—finally realized, this time in the form of Apple and Google, or iPhone/iOS and Android. (As I noted earlier, the important caveat here is that *global* excludes the important market of China, in the case of Google; there Google services are banned so a local form of Android not linked to the Google ecosystem flourishes.) Their respective app stores mean that a local app can now reach a global user. The era of contents space dominated by national telecoms is over, and the era of transnational contents diffusion is here.

There are strict limits to this global reach, of course, not the least of which is the difficulty in getting attention in an app economy increasingly dominated by incumbents who have first mover advantage. Mobile game developers in particular compete within a blockbuster economy of massive hits (while smaller games by lesser-known companies remain unknown).[3] There are also limitations, imposed by the app stores and by the local companies with which the app has tie-ups, which presents hurdles to a global release, making one national version of a given app more content rich than another—a new iteration of region locking and geoblocking on which Ramón Lobato and James Meese focus our attention, with contents and services restricted to particular markets.[4] And yet the general pivot from a world dominated by national telecommunications companies and national media corporations to one mediated by multinational platforms in which apps compete for market share in numerous nations and regions requires a shift in analytical attention. As a result, analyses of both platforms and the contents market must shift from a national to a transnational perspective.

National app and mobile developers now have the chance to have regional and global reach, uneven as this reach may be. Local hegemons have emerged, not only from the highly regulated space of China—which famously sets high barriers to or simply denies market entry for platform imperialist powers—but also Japan and Korea. Hence, by way of conclusion to this book, I would like to inquire about the place of the nation and East Asia as region in an era of platforms, through an analysis of the rise of regional chat apps and their platformization. Exploring these issues of the national and the regional in relation to the platformization of the chat app in East Asia will also allow us to see more clearly beyond the horizon of one platform era coming to an end (Japan's i-mode and the era of national carriers), the domination of another (Google's Android and Apple's iOS), and the potential beginnings of yet another (chat apps' platformization).

This conclusion treats the three most significant chat apps in East Asia: Korea's KakaoTalk, Japan's LINE, and China's WeChat. It considers them together, as platforms, in the context of questions around regional platform production, media regionalization, and platformization through the lens of the app. Addressed here are, among other things, the conditions under which a platform exceeds the nation, becomes supranational, or, in the cases discussed here, becomes regional. If chat apps are regional, the question arises as to how their regionality is produced, and what the nature of app-induced regions are. Given that these are also bound up in processes of platformization and portalization, these chat apps also foreground a return to the portal-based platform concept first developed under the stewardship of Docomo's i-mode.

On the Region and the Regional

Before taking on the apps themselves, I pause for a moment to take stock of existing theorizations of the region in East Asia. The status of East Asia as a region is particularly contentious, in no small part because of the mobilization of an idea of Asia as target within Japan's colonial empire, and the subsequent status of area studies (and East Asia in particular) within U.S.-led Cold War geopolitics of knowledge production.[5] Knowledge about Asia has always been yoked to the geopolitical drive for domination.

Despite this, or rather because of this, there has been a long, critical reckoning with the status of the region itself from within critical area studies, many accounts of which detach the regional from the mere geographical, emphasizing that geography does not region-create. I would like to sketch out at least four possible positions toward the question of East Asia as region, before moving on to the ways Chinese, Japanese, and South Korean chat apps themselves engage conceptions of the region.

A first position is staked out by Leo Ching in an influential article published in 2000, "Globalizing the Regional, Regionalizing the Global: Mass Culture and Asianism in the Age of Late Capital." Writing in the context of the regional popularity of Japanese dramas and animations, he draws parallels between a Japanese nationalist rhetoric of Asia from the period of imperial expansionism and the current moment of his writing. Yet he also notes differences, pointing to how Asia comes to be imagined by way of mass culture, which is to say, via the commodity form. Asia as regional formation, Ching argues, is indissociable from capitalist processes of commodification. Indeed, the most fundamental point Ching makes is that the regional is not a counterforce to the global or the national. Rather, the region emerges as an attempt to mediate emergent contradictions between the nation form and global capital. "Although regionalism may at times appear to oppose globalism," Ching writes, "the regionalist imaginary is fundamentally complicit with the globalist project."[6] The region here is ultimately not a fixed cultural or geographical entity but a geopolitical entity that mediates the contradictions between "the immanent logic of capital and the historical formation of nationalized economies."[7] Instead of asking, then, how the regional is opposed to the global, Ching asks why "regionalism is essential to globalism."[8] Capitalism's immanent tendency is toward transnational movement; it is complemented, historically, by the nation form. The region and regionalism "constitute a temporary mediation between the abstract logic of capital and the work of nation-states in the world economy."[9] In the particular case of Japanese mass culture's consumption in Asia, Ching writes that "mass cultural Asianism mediates between the process of the globalization of capital and the anxiety over the erosion of the nation form."[10] For Ching, then, Asia is not a geographic entity; it is a market construction, with Asianness itself being a commodity to be consumed.[11]

A second position is developed by JungBong Choi, as a rejoinder to Ching's emphasis on the economic. Choi's account is written at another moment in the pop cultural flows within Asia: the so-called Hallyu or Korean Wave that saw Korean film, dramas, and K-pop music sweep through Asia and fundamentally reconfigure cultural attitudes toward Korea in

Japan and China in particular, and the wider East and Southeast Asia region in general. As the title of Choi's article suggests—"Of the East Asian Cultural Sphere: Theorizing Cultural Regionalization"—Choi repositions East Asia as a cultural formation, as a product of "cultural regionalization." The prime mover in this case is the Korean Wave, and the fundamental reconfiguration of affective relationships within and between Asian nations because of this phenomenon. For Choi, the region is not simply an economic mediator between national and transnational; it is an accretion of historical and cultural ties that gives rise to a sense of togetherness. To think the regional requires, Choi writes, "careful reconsideration of the issues that had been kept at bay under the sway of the conceptual panoply of globalization: the unmitigated significance of geography, history, ethnic intimacy, cultural proximity, and emotional/esthetic immediacy."[12] To a regional theory axed on its place as a mediator between local and global capitalism—Ching's position on the regional, which Choi characterizes as "a mini version of cultural globalization"—Choi counterposes a theory of the region that emphasizes the importance of the proximate, the intimate, the historical ties and frictions, the wounds that bind, and East Asia's "emotional stickiness."[13] That is, the dangers of a reliance on common history, culture, or shared values that Ching isolates and writes against is precisely what Choi cautiously embraces in his outline of a theory of the regional that is bottom-up rather than top-down.[14] The Korean Wave here becomes the emblem of the creation of an "East Asian cultural sphere."

That this cultural sphere is created in part because of a "subtle embedding of traditional values" as being a "point of attraction of regional audiences" is both the provocation of Choi's argument and a potential difficulty, insofar as it courts culturalist explanations of the popularity of Hallyu (such as a reliance on Confucianism, which may or may not be a partial explanation for the success of K-dramas).[15] Thomas Lamarre's work on the region in his 2015 article, "Regional TV: Affective Media Geographies," attempts to square the circle, ultimately by arguing that the circulation of cultural products precedes the formation of a cultural sphere; regional circulation precedes (and conditions the existence of) cultural regionalization. The benefit here is that the regional distribution of media is meant to explain regional reception and cultural regionalization. The construction of a "feeling in common" that is the regional happens *through* media circulation rather than preceding it. Distribution, here, is key. Lamarre argues that the

> production of distribution holds together transmedial and transnational processes, . . . [which produces]—a feeling of something coming into common, of a region in common. The resulting "Asia" does not correspond with received territories and geographies but entails a sense of affective possession, emerging in conjunction with the mapping of the transmedial onto a geopolitical domain.[16]

In a word, the infrastructure of distribution is always in need of contents, and distribution channels exceed content availability. The supply of content is hence sought for—and found in Japanese anime or Korean dramas. Asia, as region, emerges as an after-effect (or affect-effect) of the consumption of media. The feeling of being-in-common emerges as a media effect of the consumption of the same content. Unlike Ching, Lamarre does not view this as inherently tied to capitalism (or as a mediation of global and national formations); and unlike Choi he does not view the patterns of circulation or reception as having been preconditioned by what came before (historical ties, emotional stickiness, etc.). For better or worse, in Lamarre's account the region emerges as an after-effect of a shared contents milieu, the sharing of contents itself being driven by distribution companies (TV stations, etc.) hunting for contents. (Why they pick *that* content may become a question in need of an explanation.)

The regional is medial; there are a "series of correspondences between the medial and the geopolitical, between media and nations."[17] This Lamarre refers to as "*media process geographies.*"[18] The account of the region must be rerouted, then, through media technologies rather than shared values or shared predispositions toward particular kinds of content. Regions are formed after media circulation, not before. The upside is that this argument suggests regions are prospectively formed, created as media effects. The downside, though, is that it does not explain how the region itself could be a presupposition of media circulation in production; nor does it examine the stickiness of the longer historical ties (and presumably distribution circuits) that Choi points to.

A final intervention into debates around the region that I wish to note comes in the form of Stephanie DeBoer's important book on the regional coproduction of films, particularly from the 1960s onward between Japan and China, or Hong Kong and Taiwan, depending on the period in question. DeBoer usefully shifts our focus from reception (Ching and Choi) and distribution (Lamarre) to production, the last being her point of emphasis. DeBoer situates coproduction as a means of producing the region: "Co-production's significance also lay in its function as a technology— that is, as a mode of production that potentiated new forms of encounter, expression, and, ultimately, identity for the region."[19] DeBoer's attention to coproduction also challenges the sequencing of the national to regional to global, suggesting the importance of "seeing and thus interrogating alternative cultural production or directions of flow."[20] Finally, her account asks us to shift our attention to the realm of production as the site from which to think processes of regionalization.

I draw attention to these crucial debates around the question of regionalism in and around East Asia in order to briefly suggest the diversity of positions (and there are of course many other accounts left out here), and the complexity of the question of what makes a region. Here I engage regionalism in a preliminary manner, working through the ways that the very similar chat apps—KakaoTalk, LINE, and WeChat—are generative of or "coproduce" a sense of the regional (in their reception, distribution, and production), and draw on a preexisting set of cultural tendencies within the region (such as the dominance of a character-centered mass culture). While this account is necessarily brief, the endgame is to offer a brief account of a different scale at which to position the development of mobile media post-i-mode, and to interrogate whether the regional sense generated by the platformized chat apps here constitutes something like a resistance to the global operating systems of Android and iOS (in the manner of Choi), or whether it is (more in line with Ching) simply their regional corollary.

Regional Apps

The study of apps is rife with methodological challenges, from the constant iterations and updates, to the geoblocking, geolocking, or geofencing of features to specific areas and versions of the apps.[21] Simply accessing the feature-rich Japanese version of LINE on a phone located outside Japan takes some effort—more, in this case, than a virtual private network (VPN). In media times past, region locking presented some trouble for accessing DVDs not released within one's own region. Access to apps and features not within one's own region presents a whole other level of difficulty, from the need for a VPN to the creation of Japanese accounts for Facebook and Google. Even if one can access the app, the changeability of the interface, depending on the build, and the pace at which features are retired and new iterations are released throws up challenges for the study of these new and highly unstable media objects.

Apps are, in short, unstable objects that throw up challenges for both media studies and area studies. Ben Light, Jean Burgess, and Stefanie Duguay have recently offered a lifeline in a suggestive article on the challenges and methods for studying the app. Among other things, they note that "apps' technical closure presents empirical challenges to digital media researchers."[22] This is all the more the case when these apps are located in a particular geographical milieu in which features are locked down or restricted; as Benjamin Burroughs and Adam Rugg observe, "The Internet is increasingly being regulated, controlled, and restricted at the national level."[23] This is all the more true at the level of the smartphone, whose geographical mobility is bound by a correspondingly rigid lock-in to ecosystem and national boundaries (generally tied to one's IP address and the address associated with one's app store). The Google Play store may be more or less global in reach, but each iteration of the store is rigidly nation or region specific, meaning that the app in question may not be available, or may have limited features, if it is available at all. The Japanese version of LINE is feature-rich, including a ride-hailing service, food delivery, and even a shopping service, music streaming, and so forth; the Canadian version of LINE is far more restricted, offering basic chat services and some social media timeline features but not much else; in China, the app is blocked entirely. None of these obstacles are insurmountable, but they do present significant challenges to the study of the app as a unified entity, or a coherent experience. Given these challenges to the study of apps, I adopt a variant to the "walkthrough method" that Light, Burgess, and Duguay propose, with the caveat that while I have access to the Japanese features of LINE, I am using the Canadian versions of Kakao-Talk and WeChat.

If smartphones are now at the center of digital life, these chat apps are increasingly at the center of their users' digital lives. They are the medium of communication—even that communication for communication's sake, or pure communicability—they are sources and disseminators of news and culture; and they are the mediators of commercial transactions of all kinds, from digital currency to Uber-like taxi services. Each have market dominance within their respective countries, but they are also actively borrowing from each other, building on what the other has done. Nowhere is this more evident than in their use of stickers, drawing on a regional visual culture, and leading to an understanding of the chat app itself as a form of regional coproduction.

In their discussion of KakaoTalk, Dal Yong Jin and Kyong Yoon persuasively argue, "Korea serves as a curious test bed for the future of smartphone technology, as Japan did with the development of mobile phones, known as keitai, in the early 2000s."[24] In fact, these chat apps turned platforms are the inheritors of Japan's feature phone / *keitai* platform legacy. Arguably along with Kakao, we can include LINE and WeChat, and safely say that these East Asian countries are now the test bed for the platformization of the chat app.[25] This in turn demands a reflection of the relation between region and media in the context of the chat app. While this isolation of the chat app within the context of a discourse of futurity falls into the category of a futurist "technoregionalism" that DeBoer rightly critiques in *Coproducing Asia,* in what follows I examine this futurist regional imaginary, including its actual effects on what I call the noncollaborative coproduction of these chat apps.[26] In examining noncollaborative coproduction I take up DeBoer's emphasis on filmic coproduction as the production of the region but shift it to the analysis of the development of apps themselves. These apps are not coproduced in the strict sense; they are not cofinanced and often are not jointly produced (LINE being a significant exception here, as a coproduction of a Korean company with its Japanese subsidiary). And yet—like much in the tech world, but particularly evident in the feature-rich chat app ecology—app development in East Asia is fundamentally in dialogue, or in mutual replication and copycatting, such that we can say they are iterations of each other. They copy features, interface, and offerings; sometimes they even have identical sticker sets. They also have almost simultaneously transformed themselves into more than chat apps, becoming portals to digital lives. KakaoTalk, LINE, and WeChat are the major chat apps of each country, and they emerge at a moment when chat apps move from being useful ways to talk to friends and family near and far to becoming platforms unto themselves.

If East Asian chat apps have shown themselves particularly ahead of the curve in their platformization, it is for their feature-rich environments that obviate the need to ever leave the apps (or the branded suite of apps, as the case may be). In addition to the chat function (enhanced by the inclusion of a vibrant visual culture of stickers, which I discuss below), as well as multiperson phone conversations, and social media extensions of these (from timeline, to broadcasting, to photo-sharing), one can use the apps to pay for things in the real world, shop for clothes via the app, stream music, read the news, play games, order food, reserve taxis or ride-sharing equivalents, and, for WeChat in particular, settle financial transactions, do one's banking, order plane tickets—and the list goes on. These chat apps displace the idea of having one app for each function (the "one app = one use" principle), introducing multiple functions into one app; the app functions as a smartphone-within-the-smartphone, a meta-smartphone or catch-all app that replaces the smartphone OS as the hub of digital life, and has the potential to become a megaplatform like i-mode. The chat app becomes the portal for all things digital.

As the meta-smartphone, catch-all app that replaces the smartphone OS as the hub of digital life, these apps appear as potential challengers to iOS and Android, even as they run on them. They suggest that a shift may occur wherein the power centers of the smartphone experience may be shifting from the operating system to the chat app as meta-smartphone and megaplatform. This transformation of chat apps into platforms also demands that we interrogate the horizon of the economic domination of U.S. platforms that Dal Yong Jin terms "platform imperialism." The chat apps are dependent on the dominance of iOS and Android on the one hand, and on the other hand they build something nationally or regionally spe-cific on top of these smartphone systems, which may eventually supplant them. That is to say, if ever there were threats to the current smartphone market domination by Google and Apple, they may very well come from the platformization of the chat app, wherein the chat app slowly accrues all the capabilities of the phone—from phone to social media to banking and games—potentially displacing the smartphone operators Apple and Google in the same way that smartphone operators displaced the dominance of the telecoms starting around 2007–8.

Platformization seems to proceed in tandem with portalization: the apps become portals to entire worlds of activity, resuscitating the portal concept. At the very least, this terminological modification is how this transformation is being explained. Take LINE, for instance. At the 2014 annual LINE conference (geared toward developers and business partners, and broadcast on Ustream), the keyword of the event was *platform*. The app was presented as a platform for sociality, for communication, for commerce, and the conference was organized around the "possibilities to further evolve the platform"—in the words of then CEO Morikawa Akira. LINE was to be both a "life platform" and an "entertainment platform," marking a shift toward the increasing positioning of the app as more than a communicational device, as something integrated into daily life.[27] By the 2017 LINE conference, the slogan had become "Close the Distance"—as LINE's corporate mission—and one of the key introductions was the term *portal* (see Figure 11). As Inagaki Ayumi, head of the LINE planning department, explained, the news tab included on the upper-right side of the Japanese app would be replaced by a portal tab, which would become a gateway to content and services, such as train information, fortune telling, weather, and the news. This portal will also be the access point to contents, such as manga, music, and video, which are currently housed in their own separate apps (see Figure 12).[28]

This reconfiguration of LINE as portal echoes Korean Japanese journalists Shin Mukoeng and Ha Jonggi's critical account of LINE in their 2015 book, *Yabai LINE* (Dangerous LINE), where they write: "LINE is no longer a simple messenger app, it has already attained a new form of portal media specific to the smartphone."[29] This transformation recalls the portal-based interface of i-mode, wherein one moves from a landing page to other contents and services on offer. The platformization of LINE would

Figure 11. Screenshot from LINE conference, Tokyo, 2017.

Figure 12. Screenshot from LINE conference, Tokyo, 2017.

seem, then, to require the portalization of LINE, turning it from a chat app with a series of related apps (music, games, manga, etc.) into a centralized hub for the user's media life.[30]

LINE, KakaoTalk, WeChat: The Platformization of the Chat App

The prototype for LINE—and likely WeChat and Kakao to some degree as well—was Docomo's i-mode, which as we have seen rolled out to proto-smartphones (feature phones) in 1999, monopolizing the market until the mid-2010s.[31] What Docomo did not successfully do, however, was break out of the nation. Despite attempts at expansion, its market remained Japan.

The entrance into Japan of the iPhone in 2008 was what broke the stranglehold of the telecoms on the contents market and weakened Docomo's platform power. Japan's then-smallest telecom, SoftBank, signed an exclusive contract to sell the iPhone, much like AT&T did in the United States. Docomo and au responded by turning to Android as a way to counter the draw of iPhone. If the smartphone explosion by then seemed inevitable, it also transformed the formerly all-powerful telecoms who were the brokers of transactions and platform powers into mere carriers of data and communications. Apple's App Store and Google's Google Play became the platform powers that Docomo, au, and SoftBank once were. What was different this time around is that they had a de facto global reach. Or almost global: Google Play does not have much of a presence in China, where it is formally banned, and many other stores have developed in its place.

It is in this environment of power shift from telecom to OS provider as platform that we find the emergence of chat apps, and their eventual platformization. As i-mode creator Natsuno Takeshi suggests, these apps would have probably never emerged if they had been developed during the reign of the telecoms—at least in Japan.[32] Chat apps that ate into i-mode's own communicational services would not have made it past the inspection process charged with deciding what became an official site or service and what did not. In the smartphone era, however, Google and Apple care more about users and developers than carriers, allowing for the emergence of services that directly contravene the profit sources of the carriers—phone lines and their use, and texting (they still, however, require a phone number to initiate). In Japan, LINE has even launched its own branded, low-price mobile data service (a so-called mobile virtual network operator [MVNO] that leases data capacity from larger telecoms), symbolically taking the place of the carrier itself. LINE would have been a direct challenge to the grip on messaging and talk functions that the telecoms had in Japan, just as KakaoTalk would have likely been seen as a threat to Korean telecoms' grip on texting.

Native to the smartphone era, these apps quickly establish themselves as the de facto means of communicating within national and transnational diasporic communities; they also seek to expand regionally within Asia, with their eye to eventual global expansion. Much as platforms such as i-mode created the environment for platformization of GREE, Mobage, and Dwango, the OS platforms of iPhone and Android create the environment for the platformization of the chat app. Moreover, these apps themselves are situated within a regional visual culture, trading off and emulating one another's features. In what follows, I will briefly introduce each app, compare their features, and then move on to some preliminary conclusions about how these apps collectively figure the regional.

KakaoTalk was the first chat app to launch in the region. It launched in March 2010 and currently dominates the South Korean market, installed on 93 percent of smartphones and having 48 million monthly active users, 40 million of which are in Korea. Their second-largest market is Indonesia. In 2014 Kakao merged with Daum, a portal site that is the fourth-most-visited site in Korea.[33] According to Jin and Yoon, "The emergence of the KakaoTalk-scape has been one of the most noteworthy media phenomena in Korea since the introduction of smartphones."[34] In *Yabai LINE,* Shin and Ha argue that the war of the chat apps that starts with the emergence of Kakao should in fact be historicized as the second generation of messaging software. The first generation emerged with the PC-based MSN messenger, which started in 1999 and began to take off in Korea in 2000. Around that time, competing Korean messaging services also emerged. While MSN had the largest market share at first, around 2004 it started to be eclipsed by Korean chat services like NateOn and Daum Messenger.[35]

LINE was developed by Naver Japan, South Korean internet search giant company Naver's Japanese subsidiary. Naver is the number 1 page and search engine in South Korea, with approximately 81 percent of the PC search market in 2014 and 77 percent in 2015—one of the few countries not to be dominated by Google.[36] Naver is a portal, along the lines of Yahoo! (which, as Yahoo! Japan, owned mostly by internet entrepreneur Son Masayoshi, was for a long time the number 1 site in Japan). Naver established a Japanese subsidiary, NHN Japan, in 2000. NHN Japan bought out Livedoor, a once mighty search engine and internet service provider, in 2010.[37] NHN Japan aimed to replicate the success of Naver in Korea but failed to capture much of the search market in Japan. What made its name was the release of LINE in June 2011. Stickers or stamps, the large emoticon-like objects that mark the visual culture of LINE, launched several months later, in October 2011. The official narrative of LINE is that it was created and launched immediately after the March 11, 2011, earthquake-tsunami-nuclear disaster in Japan (known as 3.11), when low bandwidth data-based communication tools such as Twitter enabled families to communicate with each other in the midst of the disaster. Nuancing this post-3.11 narrative somewhat, Shin and Ha show that a chat app was already in the works at NHN Japan at the time. More important still, they demonstrate that LINE is a version of the chat app first developed at Naver Korea as an unsuccessful attempt to take on KakaoTalk. The 3.11 disaster lent added urgency to the project and gained the parent company's firm backing.[38] As such, though LINE is presented as a Japanese company within Japan, it is better to regard it as a collaborative project between Japanese and South Korean engineers and designers, one that is fundamentally influenced by the i-mode model of platform building on the one hand, and the Kakao-Talk example of the chat app on the other. Indeed, just as LINE is seen as Japanese in Japan, it is regarded as a native Korean chat app in Korea. LINE is currently the most popular chat app in Japan, and it is also counted as the second-most-used social media platform there (with 54 percent market share), trailing only YouTube.[39] It is the dominant messaging app in Thailand and Indonesia, and competes in the Korean, Malaysian, and Mexican markets.[40] LINE operates as a suite of services, with many features built into the single chat app and other LINE-branded or LINE family services available for download, which open upon launching them from the main LINE app. The LINE chat app hence becomes the portal or hub for all smartphone life, displacing the smartphone OS as the ground upon which the smartphone experience is built.

WeChat was launched by Chinese internet giant Tencent in March 2011. Tencent actually got its start as a PC messaging service, QQ, in February 1999 before turning itself into a web portal giant in 2003 (reaching the status of the number 1 web portal in 2006).[41] WeChat has 570 million monthly active users, 100 million of which are outside the country (meaning a user base of 470 million in China), making it the largest chat app in Asia and the third-largest chat app in the world.[42] With its roots in QQ, WeChat has been part of two generations of chat applications: the first era of PC-based and the second era of smartphone-based chat applications. It also shows the revival of the web portal model in the app age. WeChat is the most feature-rich of the chat apps mentioned here, and it is one from which LINE and KakaoTalk clearly take inspiration. Or perhaps it is better described as a mutual borrowing.

All three apps are platforms in the three senses used in this book. First, they position themselves as platforms for exchange, from stickers to cabs to food delivery, mediating third-party transactions and providing a medium of settling these transactions. All three aim to be the means of settling financial transactions in particular, with WeChat leading the way on this. They are thus the typical *mediation platform*. With varying and indeed dizzying degrees of complexity, they mediate forms of transaction between users and an increasingly large number of third parties—from financial institutions, to news agencies, to advertisers, to goods sellers of all kinds. They are digital marketplaces and matchmakers. Second, they all function as *technology platforms*—in the sense of a technological foundation or base—on which to build gaming empires, which make up a substantial part of their income stream. In this second manner, they reinvent the portal, becoming gateways to a world of social games, many of which the companies create themselves. Third and finally, they participate in an economy of the exchange of so-called stickers or stamps, becoming something of a communicational and expressive *contents platform* in so doing. Communication is the keyword for the three app companies, yet this communication is not simply linguistic but deeply visual, firmly bound up with the vibrant character and celebrity-driven visual culture of East Asia. (Of course, we should add the caveat that while character culture is particularly strong in East and Southeast Asia, it is also a global culture, as Christine Yano shows in her analysis of Hello Kitty's travels across the Pacific.[43])

In its early days, the visual culture of stickers was the most defining feature of LINE, and indeed its proprietary characters continue to be the face of the brand. Even as LINE now expands into home digital assistants, the second generation of the voice-based assistant (much like Google Home or Amazon's Alexa) promises to come in the shape of its signature characters, Brown and Sally. LINE introduced stickers in October 2011, KakaoTalk in 2012, and WeChat in 2013.[44] The initial concept for the LINE stickers—early on the most immediately remarkable feature of the app—

likely comes from a similar concept developed in the context of Docomo's i-mode system, where there were larger emoji for purchase called "deco-mail," which Natsuno suggests were the visual precursor to LINE's stickers, an assertion with which LINE's manager of the sticker business planning team, Watanabe Naotomo, agrees.[45] In their scholarly study of the use of LINE stickers, Kana Ohashi, Fumitoshi Kato, and Larissa Hjorth also emphasize the contiguity between LINE's stickers and i-mode's deco-mail.[46] But despite the initial inspiration from i-mode, the extent and breadth of visual expression within LINE stickers is stunning and innovative. There are four categories of stickers on LINE, which also demonstrates the sticker stop itself operating as a multisided platform. First, there are free LINE stickers, generally of its signature characters—Brown, Cony, Boss, and so on—designed in-house and advertised as "Asia's number one character."[47] There is now a whole business built out of these characters, with goods for sale in physical LINE stores in Tokyo, Seoul, Shanghai, New York, and elsewhere, emulating the merchandising strategies of giants like Sanrio.[48] These characters have also become corporate mascots, with oversized characters adorning its Tokyo headquarters. Indeed, size and scalability have become some of the defining ways the characters have been used within store space to attract customers for a unique retail experience—such as the 3.2-meter-high "Mega Brown" plush in Seoul's Line Friends flagship store in the Myeongdong district. Second, there are LINE stickers for sale that are made by LINE, often based on popular existing characters or stars (whether music, television, or film stars). These are tailored regionally, sometimes for IP reasons, sometimes to appeal to local design preferences (likes and dislikes) based on sticker usage statistics.[49] LINE tailors character design to local or national tastes, and to monitor character sticker use means that we must acknowledge the emergence of something akin to "big character" alongside "big data"—that is, data-driven character creation.[50] Increasingly, design decisions will be made based on use data analytics—and LINE is clearly at the forefront of this transformation. Third, there are stickers for sale on LINE Creator Market, where amateurs ("creators") can sell their own creations as sticker packs. Some creators have become famously rich, and quit their jobs because of their LINE income stream, although it bears noting that LINE extracts a steep 70 percent cut from all sales (compare this to Docomo i-mode's 9 percent, or Apple and Google's

30 percent). Fourth and finally, there are promotional stickers, often available for free, that nonetheless bring in revenue for LINE as a kind of advertising where users themselves seek out the ads. For instance, KFC Japan may make stickers of Colonel Sanders that it then pays a steep fee for LINE to distribute via its platform.

Stickers can be gifted and sent but cannot be used outside the app, ensuring that the images remain within the LINE app and attached to the particular LINE account. Stickers are hence an important income stream for LINE. The breakdown of the LINE revenue stream for 2013 was 20 percent stickers or stamps and 60 percent for games or in-game items in 2013 (with official account fees and advertising stickers accounting for the remaining 20 percent); the 2015 figures were 41 percent of earnings from games and other contents, 24 percent from sticker sales, 30 percent from ads, and the remaining 5 percent from the smaller e-commerce initiatives (including LINE Pay and LINE Taxi).[51] By 2017 ad revenue had jumped to 45.5 percent of all revenues, with sticker sales taking up a relatively smaller 18 percent (though some sticker initiatives may be calculated as ad revenues—as in the KFC example). This more-diversified revenue stream represents the significant increase from LINE's advertising initiatives, which includes official corporate accounts, for which the platform charges steeply, and sponsor-paid stickers, which it calculates separately under advertising revenues. LINE counts its four major markets as Japan, Taiwan, Thailand, and Indonesia.[52]

KakaoTalk comes with a set of free stickers but also operates a sticker store, where sticker packs cost around two dollars. KakaoTalk's stickers are unique for their tie-ups with celebrities, alongside the character-based stickers (though here too the interregional copying by the chat apps means that LINE also has celebrity stickers, including the popular K-pop group BTS).[53] WeChat, by contrast, recently launched a sticker store but operates as a much more open platform; it does not charge for the stickers that are available, and most people create their own stickers or GIFs, or download those sent by others. This porousness of the app to visual culture outside the app is something that marks WeChat in particular. That said, there are also stickers that draw on the comics and manga visual culture of

East Asia. For instance, we find stickers of the iconic manga and anime character Astro Boy in all three apps. No doubt the cultural history or habituation to paying for content fostered in Japan under the feature phone system prepared Japanese users to pay for stickers; likely the widespread gaming culture in Korea with its associated freemium model of paid in-game items prepared Korean users to pay for some of their stickers; in China, the model of a free internet has moved from PC to the smartphone—and so we find a model where stickers are free (WeChat makes most of its income from its associated payment system; popular sticker makers in turn parlay their WeChat fame into merchandising and design opportunities).

In sum, all three apps exhibit a similar convergence around platform features and contents on offer, including a portal-style visual interface that can be personalized and is deeply integrated into the visual culture of East Asia and its contents industries through its character- and celebrity-oriented sticker communications model.

National Apps, Regional Scale

LINE is blocked from the mainland Chinese market, where WeChat is the dominant app. KakaoTalk has first mover advantage in Korea and overwhelming market share. LINE securely dominates the Japanese chat market. All three have plans for expansion and have made inroads in other markets as well, but their main site for services and focus remains the nations where they dominate market share. This national expansion and national domination can be explained in various ways, including first mover advantage—an economic concept that states that the first to a market can gain overwhelming market share there, such that the newcomer will not be able to break in—and network effects, which ensures a winner-take-all tendency in the tech world.[54] Beyond first mover advantage and network effects, however, the success of Kakaotalk, LINE, and WeChat respectively depends on their canny local tie-ups, media consolidation, and crucial partnerships that enhance their ability to provide services in their countries of origin, even as the absence of these may have limited their expansion beyond their national boundaries. (Again, LINE forms a complicated third model that I will have to explore elsewhere; it is simultaneously national in two senses: Korean and Japanese, even if its Korean market share stands at around 10 percent.)

Yet even if they are national first, the conditions of the global dominance of Google and Apple in the smartphone markets also allow for the expansion of these apps into other national spheres, and into other Asian markets in particular—where the affinity for a kind of character culture makes these apps more likely to catch on there than elsewhere. Riffing on Ian Condry's framing of the character itself as a platform, we might say that the character-based stickers were a platform for the national successes and regional expansion of these chat apps.[55] This is particularly true in the case of LINE, which has built a secondary business of character merchandising and physical stores throughout Asia—even in China, where the app itself has been blocked since 2014.[56] Kakao is mimicking this character strategy itself, creating experiential retail environments where merchandise and café and gigantic 3-meter-tall characters made for visitor photo ops come together in one store. If characters can be platforms (in a rather metaphorical sense), they are also more literally the contents of the app platform; character-based stickers are the communicational *contents* of these new app platforms, in the sense of transmediated, packaged IP discussed in chapter 1.

There is also a second type of regionalism that bears consideration here: the regionalism of app production and development. The apps have developed separately but in dialogue and under conditions of mutual influence, as is evident in their many shared features, similarity of interface, and portal function. To abuse the term *coproduction* DeBoer uses in relation to film and television, we might say that these apps are coproduced, albeit not in an explicitly coordinated manner. We might call these projects "noncollaborative coproductions." A commonality in that which is produced can, in turn, generate a commonality in experience itself. Shifting back to the moment of distribution and its consequence for a media-produced sense of regionalism that Lamarre articulates, we might speculate that the chat apps, in creating a parallel experience of sticker-based communication and smartphone use, in their common form, in their similar set of affordances and communication models, may produce something like that "feeling of something coming into common" that media regionalism produces.[57] This, at least, is one thing the noncollaborative coproduction of chat apps in Asia forces us to think. Simultaneously alike in their services offered, their portal-like format, and their tripartite platformization drive, these chat apps remain national in their services and their predominantly domestic user base and transnational in their affordances and user experience. The borrowing and mimicry they all engage in, a kind of noncollaborative coproduction of apps, go beyond creating a similar product; they create a similar experience, and this similarity they create constructs a "something" in common, a regionality that may or may not be felt by the apps' users but arguably creates the basis for shared feeling and shared practice, particularly around the shared icons (characters or regional celebrities) that circulate in the form of stickers, and that produces (and perhaps draws on) the emotional stickiness of cultural regionalization that Jung-Bong Choi emphasizes in his account of regionalism. This commonality emerges from different contexts, corporations, and nations, but the commonality of experience is itself significant.

As WeChat, LINE, and Kakao expand elsewhere in Asia in particular, we are led to consider what this means for thinking platform development in a regional frame. In other words, what does it mean that apps competing for market share might in fact be producing something like a feeling in common that gives rise to the region in the first place—since their very competition is nonetheless based on what is common (the similarity of apps experience)? The similarity in their offerings and features, their similarity in development, and their similarity in visual layout and interface are striking. One tends to replicate what another has done in their home market. For instance, LINE Live takes the concept of Korea's AfreecaTV streaming service (a user-broadcast service that creates celebrities out of user broadcasters or "icons"), using its monetization model based on fans giving paid stickers to the broadcasters (itself based on the earlier Korean social media platform Cyworld), as well as Niconico Video's comments overlay. While just one example, LINE Live shows the way a service in one market will quickly be deployed in another adjacent market, albeit with local inflections. These apps' similarity, in the combination of securing a national market with aspirations for regional and eventual global expansion, brings them together in a coherent movement toward something beyond the smartphone, and possibly something beyond Google and Apple. That is, these chat apps also paint a slightly different picture of the platform imperialism hypothesis, proposed by Jin in relation to the increasing prominence of Apple and Google as the technological and cultural mediators of everyday life around the globe. LINE, for instance, offers its selections of apps within the app itself. To download the app one must leave it, making it still reliant on Google's Google Play or Apple's App Store as the space from which to access the app—for the time being at least. And yet the payment systems and portal quality make LINE increasingly into a self-contained system, a smartphone within or on top of the smartphone, an incipient megaplatform—and similarly for KakaoTalk and WeChat. This platformization and portalization of the chat app, the process whereby the chat app becomes a megaplatform like i-mode was, suggests a possible future wherein the chat app becomes the site of accumulated value creation rather than Google and Apple. As we saw in chapter 1, the shift to contents was inspired in part by a shift from hardware manufacture to software as a site of value creation. Hardware became a business with small margins, as well as a race to the bottom in terms of price. Might a similar development away from Google and Apple be in the cards in a regional chat app future? "Might" indeed is the crucial qualifier here, but this scenario offers one possible regional counterimaginary to the current Silicon Valley dominance. And yet, even as we gesture toward this counterhegemonic possibility, we would also do well to recall the caution with which Leo Ching approached the regional: as a mediator for global capitalism rather than resistance to it. These chat apps may ultimately be mediators between nation and transnational capital, complements to Google and Apple rather than sites of resistance to them. Indeed, as this book has emphasized repeatedly, i-mode was the impetus for a model of the platform economy that it instilled in its users as habit, before this was globalized by Apple and Google. Hence a return to the region is by no means a liberation from the platform economy.

On this note of critical caution about the status of these chat apps vis-à-vis the platform economy and platform capitalism, I return for a moment to the emphasis on historical consciousness of platform development that this book has championed. These chat apps are alike in their resemblance to the platform strategies first developed for the mobile by i-mode, being semiclosed systems that provide contents (news, games, and now video streaming), information, payment services, and on- and offline commerce. They each aim to become the "Hand of the Buddha," or megaplatform that i-mode was in its heyday. LINE, KakaoTalk, and WeChat are hence an extension of i-mode's trajectory, much like iOS and Android themselves are. And so, the platform story does not end with the transition from i-mode to Android or iOS but instead continues with the transition to all-purpose chat apps, as is perhaps appropriate for the platform as concept and techno-logical system anchored around a layered structure: one system built upon another, one the base for the other's newer system, each displacing the other. This is platform capitalism, to be sure, but this system is somewhat more complicated than the contemporary mythologies of Silicon Valley's origi-nalism that we encounter again and again.[58] If this book has erred on the side of an overemphasis on East Asia and Japan in particular as a locus for platform production through a historical lens, this is conceived as a neces-sary countermeasure to the presentism of tech writing and the correspond-ing overemphasis on the United States as the site of platform production and politics. As this book has argued repeatedly, a firm critique of platform capitalism must be grounded in a better grasp of the development of plat-forms and platform theory in their historical and geographical diversity.[59]

The Futures of the Platform Economy

This returns us to the question of the place of the nation vis-à-vis plat-forms. Over the course of this book we have seen the shift from national platform (i-mode) to global platform (iOS/Android) to the rise of regional platforms in the form of chat apps, each stage of which participates in the national, global, and regional commercialization of the internet. The case of chat apps returns us to the methodological problem of where and when to begin a history of the present, and how to trace and narrate the genealogy of the platform that is at the center of this history. Chat apps, we have seen, are both novel as a technological form, as well as firmly indebted to the platform concept of i-mode—including the portal format that they increasingly adopt. As these chat apps become the newest gears in the motor of platform capitalism—with China and WeChat increasingly cited in the popular press as the newest challengers of Western tech companies in a platform capitalist iteration of the "twenty-first century is the Pacific century" argument—it is time to take the regionality of East Asian platforms seriously.[60]

This once again returns us to the impetus for this book: the need to carefully address platforms and their associated rhetoric in their national, global, and, in this emergent case, regional specificities; to treat the platform as a historically specific form; and to attend to platform capitalism's regionally specific iterations—to attend, that is, the local constructions of the platform economy and its organization of wor(l)ds.

In this spirit, *The Platform Economy* has offered an account of one of the theoretical and practical geneses of the platform—as theoretical inquiry, discursive object, business practice, management object and technique, and mechanism of engagement. It followed this genesis and its subsequent unfolding at multiple scales: transnational (Japan–U.S. circuits of knowledge exchange and discursive influence with mediating-translating agents like McKinsey & Company), national (Japanese platform development at NTT's Docomo, Dwango, and DeNA), U.S.-global or platform imperial (Google's Android and Apple's iOS), and Asian regional (Korean and Japanese LINE, Korea's KakaoTalk, China's WeChat). In a recent revision to his account of regionalism, Leo Ching has proposed the term *neo-regionalism* to account for an emergent configuration within East Asia, one in which Japan is no longer the industrial or cultural power center of the region, and where the power balance has shifted from Japan to Korea and China.[61] This era of neo-regionalism is also supposed to see the displacement of the United States as a central axis of Asian geopolitics. The same, in the rise of chat apps, may be happening in the realm of mobile media—should the *potential* displacement of Android and iOS that WeChat, KakaoTalk, and LINE seem to foreshadow in fact come to occur. The platformization of these chat apps may be one of the motors of neo-regionalism in Asia, even as they are examples of the continued pervasiveness of platform capitalist mediation.

This book rewound to a moment when looking at Japan circa the year 2000 appeared to be looking into the future of mobile media. At that time, Japanese mobile platforms were widely predicted to be the world's mobile future. This future indeed came to pass, but not in the form imagined. Our platform present looks a lot like i-mode's predicted future—with mobile-centric, platform-mediated financial transactions defining our experience of computing, and habituating us to having a paid relationship to contents and services within our everyday interactions. And yet the advertising centrism, data surveillance, and American platform dominance that also mark this present confront us with distinct analytical challenges of their own. Many journalists, scholars, and critics have taken these issues up, especially around surveillance. This book's wager has been on the importance of offering a history of the platform less told—a focus on Japan's platform history, the development of platform as keyword within managerial theory—and the analytical, theoretical, and interpretive benefits that telling such a story might bring. This shift has looked at platform managerial theory and practice as a site for seeing the overall, systemic expansion of a mediatory logic that the platform encapsulates—placing emphasis more on the systemic nature of these shifts than on given, discrete platforms. People have tended to approach platforms as objects, as things. But if this book argues anything, it is that they must also be approached as systemic transformations of market sides and society as a whole by a particular logic of mediation management.

The interaction, codependency, and management of contents and platforms that this book interrogates continue to be the burning issue of the day, as the rise of Netflix has suddenly made content creators think that—after an era of platform domination during the first two decades of the century—perhaps we once again are returning to the paradigm of the mid-1990s, wherein content is again king. The verdict is still out on that one, though, and if this book has shown anything, it is that each historically and geographically specific articulation of the contents-platform relation can be inventive and new, and that the relationship between the two is never completely settled. But one thing we can say definitively: Netflix has not succeeded by making content king; it has succeeded by deploying a platform logic that carefully manages the relationship between contents and users, to the benefit of its particular ecosystem. To recall Chris Bilton's words that I quoted in the introduction: "Attention has shifted from the *what* of content [i.e., an era of content as king] to the *how* of delivery, branding, and customer relationships—in other words, towards management."[62] The central tool and object of this management is the platform.

Given the pitfalls of future prediction and the tendency within business writing on platforms to venture into the fraught terrain of futurology, this book has studiously avoided making prognostications, preferring to show how sedimentations in discourse and habituations in practice have informed our platform capitalist present. But there is *one* prediction I am willing to make: the keywords of this book, *contents* and *platform,* will not be going away anytime soon. More important still, the managerial logics unfolding through and around the keywords and their associated practices will continue to evolve and impact our worlds. The more perspectives we have on them, and the more critical histories we have of them, the better equipped we will be to deal with the expressive potentials, managerial controls, and political challenges that platform mediations inevitably bring.

Acknowledgments

This book started off with my own sense that I needed a good answer to a pressing political, economic, and cultural question—*What is a platform?* The answers and this book developed and gained breadth through conversations with friends and colleagues near and far. In Montreal I thank first and foremost my colleagues at Concordia University, whose companionship in film studies makes my department a fantastic place to work: Luca Caminati, May Chew, Maria Corrigan, Kay Dickinson, Martin Lefebvre, Erin Manning, Rosanna Maule, Joshua Neves, Peter Rist, Katie Russell, Masha Salazkina, and Haidee Wasson. In the wider Montreal community, I thank Christine Lamarre, Michelle Cho, Charles Acland, Ara Osterweil, David Baumflek, Alanna Thain, Carrie Rentschler, Jonathan Sterne, Will Straw, Orit Halpern, Elena Razlogova, Meg Fernandes, Darren Werschler, Fenwick McKelvey, Bart Simon, Mia Consalvo, Rilla Khaled, Pippin Barr, Marta Boni, Brian Bergstrom, Alison Reiko Loader, Shira Avni, Luigi Allemano, Daniel Cross, Roy Cross, Gal Gvili, Tal Unreich, Xinyu Dong, Richard So, and Yuriko Furuhata, who have been wonderful friends and interlocutors. Michelle, Kay, Josh, and Yuri have been particularly special as co-conspirators on a long-running research collaboration and conference series on "Porting Media"—of which there have been several iterations and, I hope, more ports to come. Academic friends far flung have been equally supportive, so thankful hugs go to Tess Takahashi, Mike Zyrd, Pooja Rangan, Josh Guilford, Victor Fan, Jeff Scheible, Julie Levin

Russo, Braxton Soderman, Roxanne Carter, Michael Cowan, and Franz Hofer. It was a pleasure to reconnect with Matthew Holmes as he was beginning platform journeys of his own.

The Japan studies and Asian media studies communities in North America, Europe, and Asia have been particularly supportive of this project, which also coincided with coediting *Media Theory in Japan,* whose intellectual premise and encounters with contributors helped nourish this project. A huge thanks goes to Alexander Zahlten for his partnership on that project, and to all those involved in it. It was a pleasure to work with Jinying Li on a "Regional Platforms" special issue for *Asiascape: Digital Asia,* in which we tested some ideas presented here. Thomas Lamarre has been a continual and incredibly generous interlocutor on this and other projects. Thanks also go to Marilyn Ivy, Thomas Looser, Rey Chow, Akira Mizuta Lippit, Tomiko Yoda, Anne McKnight, Paul Roquet, Ian Condry, Weihong Bao, Rahul Mukherjee, Aaron Gerow, Hikari Hori, Michael Emmerich, Fabian Schaefer, Earl Jackson, Mitsuyo Wada-Marciano, Christophe Thouny, Diane Wei Lewis, Michael Raine, Jason Karlin, Patrick Galbraith, Anne Allison, and Andrew Campana—all of whom have heard parts of this project and offered helpful feedback.

Likewise, the media studies community has been equally supportive—especially Bhaskar Sarkar, Bishnupriya Ghosh, Dal Yong Jin, Wendy Chun, Vili Ledonvirta, David Nieborg, Anne Helmond, Nelly Yaa Pinkrah, Clemens Apprich, Barbara Klinger, Mary Ann Doane, Kristen Whissel, Karen Redrobe, Gilles Brougère, Inés de La Ville, Dominic Pettman (I'm sorry I didn't title this book *A Thousand Platforms* or *Civilization and Its Contents,* though I'll remain ever mindful of your title suggestions), Lynne Joyrich, Phil Rosen, and Ueno Toshiya.

Ramon Lobato generously read the entire manuscript and provided valuable feedback, for which I am deeply grateful. Josh Neves had an impact on the framing of this book, and I thank him for his many generative suggestions. Michelle Cho read the Conclusion and offered helpful comments and provocations.

This book benefited from the generosity and input of key interlocutors and interviewees. Ōtsuka Eiji has been an intellectual influence and interlocutor throughout, particularly as he developed his own platform critique of Kadokawa within the Japanese context. Ōtsuka was selfless in his support of my work, and I am honored to have had the chance to benefit from his guidance and companionship. I thank him in particular for his comments on an earlier version of chapter 5 published in Japanese as part of the translation and expansion of *Anime's Media Mix,* and in his critical writings on Kadokawa before and after its merger with Dwango. Ōtsuka also brokered a first meeting with Kadokawa Tsuguhiko, whose fascination about platforms in turn informs my take on the Kadokawa company and its subsequent merger with Dwango. Kadokawa Tsuguhiko and Date Yuri facilitated a first meeting with then–Dwango CEO Kawakami Nobuo, who also generously gave his time for subsequent interviews. Meeting with Natsuno Takeshi (i-mode cocreator) was in turn facilitated by Kawakami, and I am grateful to both for their time. Asano Takeshi, Sasaki Toshinao, Kimura Makoto, and Kokuryō Jirō generously made time to discuss platforms and platform theory in Japan with me. At LINE, Icho Saitō and Watanabe Naotomo were very helpful. Thanks go to Nakagawa Yuzuru for brokering a meeting with people from LINE.

Yoshimi Shun'ya kindly supported my visiting researcher position at the University of Tokyo, where I completed much of the research for the latter parts of this book. I thank him and the university for the support. Jason Karlin was also a wonderful host, even as he was leaving for his own sabbatical. An encounter with Vili Ledonvirta proved formative for this book; I thank Vili for introducing me to work on platforms from within management studies, one of the greatest gifts that led me to discover a whole parallel field of platform literature within the Japanese context, kick-started by the work of Japanese platform theorists Kokuryō Jirō, Kimura Makoto, and Negoro Tatsuyuki.

I also offer my gratitude to the media studies and animation studies communities in Japan that were particularly supportive of this project. Much of the earlier work on contents and platforms was first presented in the context of the Kadokawa / University of Tokyo Summer Program on the Media Mix in 2014. This was organized by Ishida Hidetaka and Ōtsuka Eiji, with Takinami Yuki and myself acting as coordinators. I thank them, all the students who participated in the event, and Andrea Horbinski, Samantha Close, Edmond Ernest dit Alban, Kondo Kazuto, and Suzuki Maki in particular. The animation studies community in and around Japan is incredible, and I owe these individuals a debt of gratitude for their companionship and amazing scholarship. Tanaka Yōko, Kim Joon Yang, Ishida Minori, Sano Akiko, Sugawa-Shimada Akiko, Shimizu Tomoko, Shin Gō, Tsai Chin, Stevie Suan, Asano Tatsuya, Koide Masatoshi, Kimura Tomoya, Wada Toshikatsu, and Yonemura Miyuki are wonderful friends and interlocutors. Though this project is not so much about animation, it nonetheless bears traces of conversations with them. The Japan media studies community has been equally supportive and friendly, and I thank Kitano Keisuke, Shinji Oyama, Kadobayashi Takeshi, the late Misonō Ryoko, Itō Mamoru, Ishida Hidetaka, and Yoshimi Shun'ya.

This work would not have been possible without the help of incredible research assistants. Alain Chouinard, Matthew Ogonoski, Dahye Kim, David Leblanc, and Jacqueline Ristola all provided crucial help along the way, digging up initial research on contents and platforms, sorting through thousands of entries on platforms (thank you, Alain!), and proofreading the manuscript in its final stages (for which I'm particularly grateful to David Leblanc). This work was generously supported by a Social Sciences and Humanities Research Council of Canada Insight Grant, as well as a Japan Foundation Fellowship. Thanks go to MA and PhD students past and present, Alain Chouinard, Kris Woofter, Jordan Gowanlock, Edmond Ernest dit Alban, Theo Stojanov, Jacqueline Ristola, Aurélie Petit, David Leblanc, Oslavi Linares Martinez, Colin Crawford, Zoé Gonzalez, André R. L. Petit, Jan Vykydal, and Sandy Carson. Valdis Silins has continued to be a valuable interlocutor. Thanks also to Philipp Keidl, Weixian Pan, Zach Melzer, Beatriz Bartolomé Herrera, and Joaquín Serpe, as well as all my students in my Managing Media class, taught as I was finalizing this book. Your work has sustained and inspired mine.

I owe a debt of gratitude to my editor, Danielle Kasprzak, with whom I was in conversation about this book for a long time. Ever encouraging, helpful, and thoughtful, Dani's editorial acumen has made this a better book—and a book full stop. Thanks too to Anne Carter. I extend my sincere thanks to the two anonymous readers whose critique and support allowed me to significantly improve on the initial manuscript.

To my parents, Renée and Norman, to my sister, Tara, and her family—Scott, Jack, June, and the myriad animals you keep—and to Setsuko, Emi, Daisuke, Haruto, Minato, and Rikuto: a huge thank you for your encouragement and support.

Last but never least, I give my most heartfelt thanks to my partner in life and in thought, Yuri, whose brilliance is a constant inspiration—and whose companionship is irreplaceable.

Notes

Introduction

I follow the Japanese convention of family name first, given name last, unless the scholar is based outside of Japan or has adopted the Western naming order convention in their publications. All translations are my own, unless otherwise noted.

1. Moazed and Johnson, *Modern Monopolies*, 17.

2. Parker, Van Alstyne, and Choudary, *Platform Revolution;* Srnicek, *Platform Capitalism.*

3. M. Baldwin, "Did Shakespeare Produce Good 'Content'?" Thanks to David Leblanc for this last point.

4. Barubora and Sayawaka's *Bokutachi no intānetto shi* [Our internet history] only addresses the mobile internet and the Japanese mobile in particular in the concluding 20 pages of a 230-page book—despite otherwise carefully going over developments decade by decade. This also tells something of the gendered figuration of the PC versus the mobile internet experience—with the mobile internet gendered as female, and the PC-based web as male.

5. Tim Wu's *Attention Merchants* offers another, focused not on platforms as much as around the development of attention economies.

6. The relation between words and things is most highlighted in the work of Michel Foucault, whose French title for what was translated as *The Order of Things* is the more memorable *Les mots et les choses,* or words and things. Gilles Deleuze's articulation of the relation between the two "strata" (as he would call words and things, or the seeable and the sayable) in his *Foucault* book remains an inspiration here, even if more an implicit rather than explicit one. Here I take liberty with Foucault's suggestion that discourse operates through its regularities, to make the claim that practice, too, similarly operates through particular kinds of regularities or repetitions.

7. Lazzarato, *Les revolutions du capitalisme,* 94.

8. Lazzarato, 96.

9. CEO letters to shareholders are what Ken Hyland calls "a highly rhetorical product that can have a major impact on a firm's competitive position." Hyland, "Exploring Corporate Rhetoric," 224. On financial markets and language, see Marazzi, *Capital and Language.* For an excellent treatment of platforms in the context of finance, see Vonderau's analysis of Spotify in "The Spotify Effect."

10. Kiechel, *The Lords of Strategy,* 28: "The corporation's application of sharp-penciled analytics this time not to the performance of an individual worker—how fast a person could load bars of pig iron or reset a machine—but more widely to the totality of its functions and processes."

11. Kiechel, 12.

12. Here I follow a usage of the term proposed by management scholars Martin Kenney and John Zysman in "The Rise of the Platform Economy" but also include the linguistic economy of the term *platform* rather than simply the financial or digital economy of the term as used by Kenney and Zysman.

13. David Nieborg and Thomas Poell term this "business studies"—here I refer to this body of work interchangeably as business literature or management theory. Nieborg and Poell, "The Platformization of Cultural Production," 2.

14. For a work that is timely and refreshing in its attention to language—indeed, punctuation—amid the hardware-heavy digital turn, see Golumbia, *The Cultural Logic of Computation*; and Scheible, *Digital Shift.*

15. Golumbia, *The Cultural Logic of Computation,* 1–2.

16. This book hence complements Wu's project of "making apparent the influence of economic ambition and power on how we experience our lives"—with the caveat that words and ideas help shape how the economic impacts our worlds and lives. Wu, *The Attention Merchants,* 6.

17. Gillespie, "The Politics of 'Platforms,'" 360.

18. In *Technotopia,* Clemens Apprich offers a history of the network concept and net cultures in the 1990s. While important to note the change in guard of the terms *network* and *platform,* for the remainder of this book I focus on the term and operations of *platform.*

19. Boltanski and Chiapello, *The New Spirit of Capitalism,* 84.

20. Boltanski and Chiapello, 57.

21. Lovink, "Foreword," xiii.

22. Parker, Van Alstyne, and Choudary, *Platform Revolution*; Evans and Schmalensee, *Matchmakers.*

23. The organizers of the conference were Helen Margetts, Vili Lehdonvirta, Jonathan Bright, David Sutcliffe, and Andrea Calderaro. They subsequently edited two special issues of *Policy & Internet* with the title "Platform Society." This is also the title of a book by José van Dijck, Thomas Poell, and Martijn de Waal, *The Platform Society.*

24. Liu, "Transcendental Data," 63.

25. Mailland and Driscoll, *Minitel.*

26. Steinbock, *Wireless Horizon*, 163.

27. Lamarre, *The Anime Ecology*, 162.

28. In 2000 the total mobile contents market for the proto-smartphone "feature phones" in Japan was 48 billion yen (US$442 million); it had risen to 132 billion yen in 2001, 179 billion in 2002, 260 billion in 2004, 366 billion in 2006, 472 billion in 2007, 483 billion in 2008, and 552 billion in 2009. In 2010 the market for feature phone contents was 600 billion yen, a number that shrank over the next few years as revenue shifted to the smartphone (in 2010 the smartphone only accounted for 1 percent of the contents market in Japan, however). The total market for contents across both the feature phone and smartphone expanded steadily, however, reaching 734 billion yen in 2011, 851 billion in 2012, and 1.7 trillion in 2013. Total spending on mobile contents has continued to increase, even as the share of the feature phone market decreased. Mobile Content Forum, *Keitai hakusho 2006* [Cell phone white paper 2006], 209; Mobile Content Forum, *Keitai hakusho 2011* [Cell phone white paper 2011], 150, 151; Mobile Content Forum, *Sumaho hakusho 2015* [Smartphone white paper 2015], 35.

29. Choo, "Nationalization 'Cool.'"

30. See digital strategist and consultant Bharat Anand's *The Content Trap*.

31. Bilton, "The Management of the Creative Industries," 31.

32. Goggin, *Global Mobile Media*, 4.

33. For another refreshing account of platform genesis, I once again refer the reader to Mailland and Driscoll's *Minitel*.

34. Furuhata, "Architecture as Atmospheric Media"; Gardner, "The 1970 Osaka Expo and/as Science Fiction"; Morris-Suzuki, *Beyond Computopia*.

35. On the ambience and environmentality of media in Japan, see Furuhata, "Architecture as Atmospheric Media"; Roquet, *Ambient Media*.

36. C. Johnson, *MITI and the Japanese Miracle*, 5.

37. McLelland, Yu, and Goggin, "Alternative Histories of Social Media in Japan and China," 56.

38. Jin, "The Construction of Platform Imperialism in the Globalization Era," 154. Jin elaborates the concept further in his monograph, *Digital Platforms, Imperialism, and Political Culture*.

39. Jameson, "Notes on Globalization as a Philosophical Issue," 58.

40. Jameson, 63.

41. Miller et al., *Global Hollywood*.

42. Jin, "The Construction of Platform Imperialism in the Globalization Era," 154.

43. "Dwango wa ōhaba zōshū zōeki, Kadokawa wa akagi" [Dwango expands income and profit, Kadokawa is in the red].

44. "Kadokawa-Dowango keiei tōgō" [The business integration of Kadokawa-Dwango].

45. "Kadokawa Kaichō, keiei tōgō de 'Hi no maru purattofōmu tsukuru [Chairman Kadokawa, the Merger Will "Create a Platform of the Rising Sun"].

46. Sakai, *Translation and Subjectivity.* Indeed, this is where critical area studies' caution toward national narratives and knowledge production about East Asia in general offers some help in disentangling the need to decenter the United States or the West from the attendant danger of recentering the former imperialist nation of Japan.

47. Hands, "Introduction," 1.

48. Helmond, "The Platformization of the Web."

49. Coates and Holroyd, *Japan and the Internet Revolution*; Ito, Okabe, and Matsuda, *Personal Portable Pedestrian.*

50. Ito, "Introduction," 4.

51. As Goggin puts it, "Japan pioneered mobile Internet." Goggin, *Global Mobile Media,* 30.

52. Manabe, "Going Mobile," 329. I would also like to acknowledge the crucial research of Larissa Hjorth on the study of mobile media in the Asia Pacific, in work such as *Mobile Media in the Asia-Pacific* and, with Michael Arnold, *Online@Asia-Pacific,* as well as work by Misa Matsuda, "Discourses on *Keitai* in Japan."

53. For another story, told through animation and television, see Lamarre, *The Anime Ecology.*

54. See Salazkina, "Introduction"; Steinberg and Zahlten, *Media Theory in Japan.*

55. Along with *Making Media Work, The Platform Economy* also heeds Shinji Oyama's crucial call to study the Japanese creative industries beyond those objects most associated with "Cool Japan" (anime, manga, and console games). Oyama, "Japanese Creative Industries in Globalization."

56. Holt and Perren, "Introduction"; Zahlten, *The End of Japanese Cinema.*

57. Nieborg, "Crushing Candy." I thank Vili Ledonvirta for introducing me to work on platforms from within management studies; I also benefited from his book with Edward Castronova, *Virtual Economies.*

58. Goggin and McLelland, "Internationalizing Internet Studies"; Ito, Okabe, and Matsuda, *Personal Portable Pedestrian.*

59. Cho, "Pop Cosmopolitics and K-pop Video Culture," 244.

60. Gillespie, "The Politics of 'Platforms'"; van Dijck, *The Culture of Connectivity*; Bogost and Montfort, *Racing the Beam*; Helmond, "The Platformization of the Web"; Jin, "The Construction of Platform Imperialism in the Globalization Era"; Lamarre, "Regional TV."

61. Negoro and Ajiro, "An Outlook of Platform Theory Research in Business Studies."

1. Contents Discourse

1. Anand, *The Content Trap.*

2. On the importance of walled gardens, see Gopinath, *The Ringtone Dialectic,* 41–42.

3. Once again I remind the reader that I render the Japanese term *kontentsu* as *contents,* unless this either explicitly refers to the non-Japanese context or is a

citation from an English translation of Japanese that renders the term as *content* (as translations from the Japanese to English often do). I also use the singular *content* as a translation for the Japanese *naiyō* (内容), which is the most direct translation of content or substance and is often used to gloss *kontentsu*.

4. A *hakusho,* or "white paper," is a report produced sometimes by the government or, more often in the case of Japan, by an industry-affiliated research body that provides an overview of a given industry or subsection of an industry on a yearly basis. These are published as large-format books, and are often sold at a relatively affordable price, meaning that they are fairly widely circulated.

5. See Allison, "The Cool Brand, Affective Activism and Japanese Youth"; Condry, *The Soul of Anime*; Galbraith and Karlin, *Idols and Celebrity in Japanese Media Culture*; Lamarre, *The Anime Machine*; Saito, "Magic, Shōho, and Metamorphosis." On contents, see Choo, "Nationalization 'Cool'"; Lamarre, *The Anime Ecology.*

6. I would like to take this opportunity to signal the importance of Choo's work on the governmental formulation of its contents policy, particularly around the "Cool Japan" moniker. See, for instance, Choo, "Nationalizing 'Cool.'"

7. If this chapter is a first step at delineating the wherefrom and reason for the word *contents* in Japan, it may also serve as a starting point for comparative work that would track the travels and transformations of the term *contents* in the East Asian region.

8. Ynostra, "Whose Content Is It Anyway?," 12.

9. Lev Manovich has reflected on the qualities of the computer as a metamedium. Manovich, *The Language of New Media,* 6. In his more recent volume, *Software Takes Command,* he quotes a 1984 article in which computer scientist Alan Kay articulates the computer as a metamedium: "It [a computer] is a medium that can dynamically simulate the details of any other medium, including media that cannot exist physically. It is not a tool, though it can act like many tools. It is the first *metamedium*" (105–6).

10. Frosh's work, *The Image Factory,* is particularly useful in emphasizing the homology of "information" and "content" insofar as the former "*appears* as an abstract universal, without history or geography" (196, 198). My argument in this chapter is that contra Frosh there is in fact a geographical difference to be found in the *usage* of the term *content*—as, more precisely, *kontentsu*—and that we must pay attention to this regional, national, or linguistic difference in usage.

11. Frosh, 197.

12. Liu, "Transcendental Data," 58; Kittler, *Gramophone, Film, Typewriter.*

13. "Multimedia-content business" appears in Matsumoto, "Bankrupt Media Vision Looks to Sell Software Publishing Unit," 16; "information-content business" is in Narisetti and Steinmetz, "Reed Elsevier Wins Bidding for Lexis/Nexis," A3. In 1993 the *Wall Street Journal* uses the term *content business* within a quote from the former head of the American Federal Communications Commission, Andrew Sikes. See Ferguson, "Business World," A15.

14. Levinson, "Who's Who on the Information Superhighway?," 98.

15. Gates, "Content Is King."

16. Safire, "The Summer of This Content," SM18.

17. As CEO of Intel Andy Grove says in a 1992 speech, "The content industries are the information providers" ("Need or Greed").

18. It bears noting that Japan itself played no small role in the digital shift. As writers such as Tessa Morris-Suzuki noted long ago in *Beyond Computopia*, the very conception of the information society emerged in Japan. Japanese electronics industries in turn played a major role in the proliferation of computing devices, from personal computer to workplace mechanization. See Chandler, *Inventing the Electronic Century*. The digital revolution and Japan go hand in glove.

19. On Cool Japan, see Allison, "The Cool Brand"; Choo, "Nationalization 'Cool'"; Condry, "Anime Creativity."

20. The term was coined by Douglas McGray in a now foundational article, "Japan's Gross National Cool."

21. Steinberg, *Anime's Media Mix*; Jenkins, *Convergence Culture*.

22. Azuma, *Kontentsu no shisō* [The thought of contents].

23. Nakano, *Manga shinkaron* [Theory of manga evolution].

24. Kadokawa Tsuguhiko and Katagata, *Kuraudo Jidai to "Kūru Kakumei"* [The cloud era and the "cool revolution"].

25. Satō Tatsuo, "Address to the Company," 2.

26. Hatakeyama and Masakazu, *Odoru kotentsu bjinesu no mirai* [Future of chaotic entertainment content biz], 3; Zöllner and Nakamura, *Culture and Contents*, 7.

27. Horn, *Haipātekisuto jōhō seirigaku* [Studies of hypertextual information organization].

28. Nikkei BP, "Studies of Hypertextual Information Organization."

29. Scheible, *Digital Shift*, 73–101.

30. Shimizu, *Fujitsū no maruchi media bijinesu* [Fujitsū's multimedia business].

31. Sekizawa, "Nikkei sangyō shinbun sōkan 20 shūnen tokubetsu kōgikai" [On the occasion of the 20th anniversary of the founding of Nikkei Sangyō newspaper], 5.

32. Sekizawa, 5.

33. "Kīwādo de kiku '95 jōhō tsūshin sangyō: Kontentsu" [Understanding the 1995 information communication industries by keyword: Contents], 6.

34. "Kīwādo de kiku '95 jōhō tsūshin sangyō," 6.

35. Fukutomi Tadakazu in his genealogy of the term *contents* notes that the term took root in the Japanese ministry for trade as a term geared for the industrial world. Moreover, he notes that "one also hears the term contents used in a similar context in South Korea and Taiwan, we believe in reference to the Japanese usage." Fukutomi, "Kontentsu to wa nanika" [What is contents?], 4. While Fukutomi's comment is speculative rather than definitive, it seems to be backed up by the comparative timelines of the usage of the term within South Korean and Japanese media industries; it also appears that a joint summit between South Korean and

Japanese governments is one of the first contexts for the use of the transliterated term *contents* in Korean. Thanks to Dahye Kim for this research on contents in the Korean context.

36. Shimizu, *Fujitsū no maruchi media bijinesu,* 160.

37. "Kyō no kotoba: Kontentsu" [Word of the day: Contents], 3.

38. See *Nikkei shinbun,* January 1, 4, 5, 6, 9 10, 11, and 12, 1995.

39. Noguchi, *Kontentsu bijinesu* [Contents business].

40. For instance, Ōtsuka Eiji frequently invokes *monogatari sofuto* (narrative software) in his influential book on narrative-based marketing, *Teihon monogatari shōhiron* [A theory of narrative consumption: Standard edition], 21–54.

41. Lamarre, "Anime," 24.

42. For accounts of Kadokawa, see Satō Kichinosuke, *Subete wa koko kara hajimaru* [Everything starts from here]; Steinberg, *Anime's Media Mix;* Zahlten, *The End of Japanese Cinema.*

43. Kadokawa Tsuguhiko, "Atarashii shuppan wo motomete" [In pursuit of new publishing], 6–7.

44. Satō Kichinosuke, *Subete wa koko kara hajimaru,* 147.

45. Kadokawa Shoten, *Corporate Brochure,* 18–19.

46. Maruchimedia Sofuto Shinkō Kyōkai, *Maruchimedia hakusho* [Multimedia white paper], 1993 and 1996 editions.

47. Nihon Jōhō Shori Kaihatsu Kyōkai, *Jōhōka hakusho 1993* [Informatization white paper 1993], 146.

48. Nihon Jōhō Shori Kaihatsu Kyōkai, *Jōhōka hakusho 1994,* 148, emphasis mine.

49. Nihon Jōhō Shori Kaihatsu Kyōkai, *Jōhōka hakusho 1995,* 99.

50. Nihon Jōhō Shori Kaihatsu Kyōkai, *Jōhōka hakusho 1996,* 91. In the following year, this section is renamed "Individuals and Their Lives in the Network Era." Nihon Jōhō Shori Kaihatsu Kyōkai, *Jōhōka hakusho 1997,* 105.

51. Nihon Jōhō Shori Kaihatsu Kyōkai, 176.

52. Nihon Jōhō Shori Kaihatsu Kyōkai, 194.

53. Nihon Jōhō Shori Kaihatsu Kyōkai, 194, emphasis mine.

54. Nihon Jōhō Shori Kaihatsu Kyōkai, 196.

55. Nihon Jōhō Shori Kaihatsu Kyōkai, 205.

56. Dentsū Sōken, *Jōhō media hakusho 1996* [Informational media white paper 1996], i.

57. Dentsū Sōken, i.

58. Dentsū Sōken, *Jōhō media hakusho 1998,* i.

59. Dentsū Sōken, i.

60. I draw on the Japanese usage of the term *subcultural* here, often designating a nonmainsteam or niche form of media or cultural consumption. Manga, anime, and games are often termed *subcultural media.*

61. Hatakeyama and Kubo, *Odoru kotentsu bijinesu no mirai,* 3–4.

62. Hatakeyama and Kubo, 3–4.

63. Hatakeyama and Kubo, 6–7.

64. Azuma defines contents as "manga, anime, light novels and other forms of expression that are generally called 'contents.'" See Azuma, *Kontentsu no shisō*. On the media mix, see Ito, "Technologies of the Childhood Imagination"; Steinberg, *Anime's Media Mix*; Tanaka, "Media mikkusu no sangyō‑kōzō" [The industry structure of the media mix].

65. Fujioka, *Sayōnara, taishū* [Goodbye, masses]; Inamasu, *"Posuto koseika" no jidai* [The age of post individualization]; Ivy, "Critical Texts, Mass Artifacts" and "Formations of Mass Culture"; Kohara, *Nihon māketingu shi* [Japanese marketing history]; Kojima, "Komāsharu to māketingu" [Commercials and marketing].

66. Ōtsuka, *Monogatari shōhiron* [A theory of narrative consumption]; Fukuda, *Monogatari māketingu* [Narrative marketing].

67. The following discussion is based in part on my relatively systematic review of television commercials available for viewing at Advertising Museum Tokyo, focusing particularly on the period from the 1950s through the 1990s. Print ads for this period were also examined, and demonstrate a parallel shift from need to image value.

68. Kojima, "Komāsharu to māketingu," 51.

69. Kojima, 91, 112n1.

70. Ivy, *Discourses of the Vanishing*.

71. Yoda, "Girlscape," 180–82.

72. Ivy, "Formations of Mass Culture," 253. Many Japanese economic historians date the end of postwar mass consumption to the 1973 oil crisis, but Fujioka and historians of television commercials tend to date the diversification of consumers to the beginning of the 1970s.

73. Fujioka, *Fujioka Wakao zen purodūsu* [Fujioka Wakao's productions], vol. 2, 18.

74. Fujioka, 2:19.

75. Fujioka, 2:45–46. Yoda notes that "Fujioka's notion of de-advertising echoes the iconoclastic gestures that thrived in the US advertising industry of the 1960s, referred to as 'anti-advertising' by Thomas Frank in his book *The Conquest of Cool*." Yoda, "Girlscape," 181.

76. Ivy, "Critical Texts, Mass Artifacts," 37.

77. See Ivy; Sasaki Atsushi, *Nippon no shisō* [Japanese thought]. See also Zahlten, "1980s Nyū Aca."

78. Inamasu, *"Posuto koseika' no jidai*, 30. The article from which the statement comes was originally published in 1990. Inamasu is acknowledged as one of the early thinkers to shine a spotlight on the issue of the rebirth of narrative. See Fukuda, *Monogatari māketingu*, 196–97.

79. Ōtsuka notes the interest among Dentsū admen in "story marketing" by the late 1980s. See Ōtsuka, *Teihon monogatari shōhiron* [A theory of narrative consumption: Standard edition], 323; Ōtsuka, *Monogatari shōmetsu ron* [A theory of the destruction of narrative], 13–14.

80. First published as *Monogatari shōhiron*, then expanded in 2001 as *Teihon monogatari shōhiron*, and most recently expanded, re-edited, and published as *Monogatari shōhiron kai*. An excerpt of the 2001 edition is translated as "World and Variation" by Marc Steinberg.

81. Ōtsuka, *Teihon monogatari shōhiron*, 15; Ōtsuka and Azuma, *Riaru no yukue* [The whereabouts of the real].

82. Ōtsuka, *Teihon monogatari shōhiron*, 323.

83. Fukuda, *Monogatari māketingu*, 3.

84. Fukuda, 25–26.

85. Fukuda, 19.

86. Fukuda, 26.

87. Fukuda, 26–28; see also Ōtsuka, *"Otaku" no seishin-shi* [A history of the Otaku mind].

88. Fukuda, *Monogatari māketingu*, 3.

89. Yamakawa, *Jirei de wakaru monogatari māketingu* [Narrative marketing understood through examples], 203–4.

90. Arai, Fukuda, and Yamakawa, *Kontentsu māketingu* [Contents marketing].

91. Deguchi, Tanaka, and Koyama, *Kontentsu sangyōron* [On the contents industry]; Dejitaru Kontentsu Kyōkai, *Dejitaru kontentsu hakusho 2013* [Digital contents white paper 2013]; Enterbrain, *Sekai no entame gyōkai chizu 2013 nenban* [World entertainment industry map, 2013 edition]; Nakano, *Manga shinkaron*; Zöllner and Nakamura, *Culture and Contents*.

92. Lamarre, "Regional TV," 119.

93. Arai, Fukuda, and Yamakawa, *Kontentsu māketingu*, 5.

94. Arai, Fukuda, and Yamakawa, 67.

95. Arai, Fukuda, and Yamakawa, 160.

96. Arai, Fukuda, and Yamakawa, 161.

97. Nakano, *Manga shinkaron*, 18.

98. Shapiro and Varian, *Information Rules*, 21. On digital locks and rights management tools, see Gillespie, *Wired Shut*.

99. Here again I should note the importance of the parallel term *platform* in the development of technologies that support the monetization of contents. We will see this support in action in chapter 4.

100. Willis, *A Primer for Everyday Life*, 3.

101. Gopinath, *The Ringtone Dialectic*, 23.

102. Anand, *The Content Trap*.

2. Platform Typology

1. Fred Vogelstein's *Dogfight*, a popular account of the battle between Android and iOS, offers a useful account of this competition focusing on their smartphone divisions. For popular accounts of the platform strategies of Apple and Amazon, see Isaacson, *Steve Jobs*; and Stone, *The Everything Store*.

2. This is also the subject of academic study, such as the focus on platform-enabled "connected viewing," to cite the title of a recent edited collection by Holt and Sanson, *Connected Viewing*.

3. As I discuss in chapter 5, the years 2005–6 are sometimes regarded as the end of a certain era of the contents bubble, though they by no means mark the end of spending on contents; spending on digital contents continue to rise through the present day.

4. See, for instance, Gawer, "Platform Dynamics and Strategies"; Suarez and Cusumano, "The Role of Services in Platform Markets"; for critical texts, see Jin, "The Construction of Platform Imperialism in the Globalization Era"; Plantin, Edwards, Lagoze, and Sandvig, "Infrastructure Studies Meet Platform Studies in the Age of Google and Facebook"; Srnicek, *Platform Capitalism*.

5. In this article, they suggest two main divisions, with the second subdivided into two types: (1) Platform Technology Theory / Platform Components Theory; and (2) Platform Products Theory, wherein we find subdivisions (2.1) Layer-Type Platform Theory and (2.2.) Interaction-Type Platform Theory. Their typology roughly aligns with Carliss Y. Baldwin and C. Jason Woodard's typological division of platforms research into "three distinct but related fields: product development, technology strategy and industrial economics." Baldwin and Woodard, "The Architecture of Platforms," 19. Gawer, one of the most significant contributors to the "technology strategy" stream, offers a similar breakdown, rightly noting: "One could even wonder at first glance if they are discussing the same underlying phenomenon." Gawer ultimately does suggest there is a through-line, as I will also do here. Yet perhaps the most immediate takeaway from this work is its disciplinary locus: it comes out of what we might generally call business studies, or management literature, much of which offers explicit advice on how to mobilize this for business efficiencies or business strategy. Gawer, "Platform Dynamics and Strategies," 46.

6. See, for instance, Bratton, *The Stack*.

7. Grewal, *Network Power*.

8. Interview with Sasaki Toshinao, Tokyo, October 9, 2015.

9. Shimizu, *Fujitsū no saibā bijinesu* [Fujitsū's cyber business], 229.

10. Nihon Jōhō Shori Kaihatsu Kyōkai, *Jōhōka hakusho 1994*, 148.

11. An exhaustive search of the LexisNexis database reveals the following, 1985 article to be the first to unite the computational meaning of platform with IBM: "TEKTRONIX; Announces Redirection of Graphic Workstation Development."

12. Bresnahan and Greenstein, "Technological Competition and the Structure of the Computer Industry," 2.

13. Bresnahan and Greenstein, 2.

14. Bresnahan and Greenstein, 39.

15. Andreessen, "The Three Kinds of Platforms You Meet on the Internet." This and other blog posts by Andreessen are engaged with by Ian Bogost and Nick

Montfort, as well as Tarleton Gillespie, thereby impacting the take-up of the term within contemporary media platform studies.

16. The earlier blog post to which he refers is actually about Facebook, titled "Analyzing the Facebook Platform, Three Weeks In."

17. Zittrain, *The Future of the Internet and How to Stop It*, 3–5, 101.

18. Zittrain, 30.

19. On computer as metamedium, see, for instance, Manovich, *The Language of New Media*.

20. Helmond, "The Platformization of the Web."

21. Indeed, Cameran Ashraf and Luis Felipe Alvarez Leon argue that this binary of open versus closed in relation to devices fails to acknowledge that the internet—as well as its devices—is constructed to be both. They write: "The binary model of an open or closed system should be seen as part of a broader range of geopolitical and geoeconomic logics espoused by states and other actors, such as firms, who envision and construct the internet through different territorial perspectives." Ashraf and Alvarez Leon, "The Logics and Territorialities of Geoblocking," 42.

22. *Merriam Webster*, s.v. "console," accessed January 16, 2018, http://www.merriam-webster.com/dictionary/console.

23. Bogost and Monfort, "Platform Studies," 1.

24. Bogost and Monfort, 3.

25. Bogost and Monfort, *Racing the Beam*, 2.

26. Thomas Apperley and Jussi Parikka offer a very useful overview and critical engagement with this initiative in "Platform Studies' Epistemic Threshold."

27. Mailland and Driscoll's 2017 *Minitel* is the best example of a platform studies book that takes the transactional dimension and cultural dimensions of platforms most seriously.

28. Apperley and Parikka, "Platform Studies' Epistemic Threshold," 4–5.

29. Gillespie, "The Politics of 'Platforms,'" 351, cited in Apperley and Parikka, "Platform Studies' Epistemic Threshold," 5.

30. Evans, Hagiu, and Schmalensee, *Invisible Engines*, 125. See also the excellent account of Nintendo's business practices and the lockout chip in particular in Sheiff, *Game Over*, 71, 161–62.

31. Altice, *I Am Error*, 91.

32. Srnicek, *Platform Capitalism*, 16.

33. This analysis is based on exhaustive searches of the Factiva, ProQuest, Academic Search Complete, and Lexis-Nexis databases, among many others more closely related to the car industry (or production and engineering).

34. Flint and Tomarkin, "Wait 'til Next Year," 52.

35. *World Auto Trade*, 234.

36. Womack, Jones, and Roos, *The Machine That Changed the World*, 112. For another crucial account of Toyotism as introduced to an Anglophone audience, see Kenney and Florida's *Beyond Mass Production*.

37. Srnicek argues for the importance of the history of supply chain management for grasping contemporary platforms, and what he calls "lean platforms" in particular. Srnicek, *Platform Capitalism*, 90–92. For a fascinating account of logistics, see Cowan, *The Deadly Life of Logistics*.

38. Nobeoka, *Muruchi purojekuto senryaku* [Multiproject strategies], 34. In the note that accompanies this explanation, the author adds that in newspaper articles this is often referred to as the "chassis," while in the automobile industry it is common to refer to it as the "platform" (34n4).

39. See Gawer's review of platform literature in "Platform Dynamics and Strategies," 46.

40. Wheelwright and Clark, "Creating Project Plans to Focus Product Development," 73.

41. According to Wheelwright and Clark, "In the computer market, IBM's PS/2 is a personal computer platform; in consumer products, Procter & Gamble's Liquid Tide is the platform for a whole line of Tide brand products"(73).

42. Suarez and Cusumano, "The Role of Services in Platform Markets," 77–78.

43. Gillespie, "The Politics of 'Platforms,'" 351.

44. Andrejevic, "Exploiting YouTube."

45. van Dijck, *The Culture of Connectivity*, 36.

46. "The World's Most Valuable Resource Is No Longer Oil, but Data."

47. Gillespie, "The Politics of 'Platforms,'" 348.

48. Gillespie, 348.

49. Nihon Jōhō Shori Kaihatsu Kyōkai, *Jōhōka hakusho 1994*, 148.

50. Dentsū Kenkyū, *Jōhō media hakusho 2010* [Informational media white paper 2010], 174. Note that these Dentsū white papers are published in January of the year in the title, meaning we should date the arrival of platform as a term to sometime in 2009.

51. Dentsū Kenkyū, *Jōhō media hakusho 2011*, 183.

52. Dentsū Kenkyū, *Jōhō media hakusho 2008*, 149.

53. Nihon Jōhō Shori Kaihatsu Kyōkai, *Jōhōka hakusho 2001*, 165.

54. Nihon Jōhō Shori Kaihatsu Kyōkai, *Jōhōka hakusho 2002*, 157.

55. See Lamarre, *The Anime Ecology*, "Platformativity," and "Regional TV."

56. Kawakami, *Suzuki-san ni mo wakaru netto no mirai* [The future of the internet that even Mr. Suzuki can understand].

57. Sasaki Toshinao, *Netto mirai chizu* [Net future map], 73–74. Kawakami was likely aware of this work, especially since Sasaki wrote a book about his Niconico Video service. An even earlier use of the term *contents platform* comes in Natsuno Takeshi's 2002 book *A la i-mode*, translated as *The i-mode Wireless Ecosystem*, 125–26.

58. Kawakami, *Suzuki-san ni mo wakaru netto no mirai*, 93.

59. Kawakami, 93.

60. Kawakami, 95.

61. Kawakami, 111.
62. van Dijck, *The Culture of Connectivity,* 29.
63. Kadokawa Tsuguhiko, *Google, Apple ni makenai chōsakuken* [Copyright laws that can't be defeated by Google and Apple], 37.
64. Schmidt and Rosenberg, *How Google Works,* 78–79.
65. Schmidt and Rosenberg, 79.
66. Srnicek, *Platform Capitalism,* 43.
67. In one of the more interesting accounts of platforms, Ian Condry in *The Soul of Anime* suggests that we should think of animation characters themselves as platforms. Here Condry moves the term even further away from the realm of computation but nonetheless draws on the sense of "the common" given to the term by YouTube and similar platforms. He writes, "One can think of platforms not only as mechanical or digital structures of conveyance but also as ways to define and organize our cultural worlds." "Characters," he continues, "offer a way to think about anime as a 'generative platform,' especially in the sense that they can exist somewhat independently of the storytelling to follow. This is important because in the case of anime, it is seldom narrative coherence—the story—that provides the link across media. Rather, the characters and the worlds provide that link." *The Soul of Anime,* 57–58.

3. Transactional Platform Theory

1. Gawer, "Platform Dynamics and Strategies," 46.
2. There are significant exceptions, of course; the works within or engaging with media studies that most directly draw on this work from economics are Helmond, "The Platformization of the Web"; Nieborg, "Crushing Candy"; Plantin, Edwards, Lagoze, and Sandvig, "Infrastructure Studies Meet Platform Studies in the Age of Google and Facebook"; and recent books within MIT Press's Platform Studies series such as Mailland and Driscoll's *Minitel.* Srnicek's *Platform Capitalism* also draws on this work to some degree.
3. Guillory, "Genesis of the Media Concept."
4. Liu, *The Laws of Cool,* 16.
5. Micklethwait and Wooldridge, *The Witch Doctors,* 45.
6. Schmidt and Rosenberg, *How Google Works,* 78–79.
7. I thank Vili Lehdonvirta for his generous introduction to this body of work.
8. Two useful accounts of network effects in particular and the platform literature in general can be found in Evans and Schmalensee, *Matchmakers,* 21–30, 208–9; and Parker, Van Alstyne, and Choudary, *Platform Revolution,* 29–31.
9. Evans and Schmalensee, *Matchmakers,* 14.
10. Parker and Van Alstyne, "Information Complements, Substitutes, and Strategic Product Design," 2.
11. Evans, Hagiu, and Schmalensee, *Invisible Engines,* 55.
12. Evans, Hagiu, and Schmalensee, 55.

13. Subsequently, authors such as Parker and Van Alstyne began to describe their object of study as a platform too. This marks a shift from their 2000 article, "Information Complements, Substitutes, and Strategic Product Design," in which they clearly address the problem of two-sided markets but, as we have seen above, do so using the language of *firms*, not *platforms* (13–15).

14. Parker and Van Alstyne, "Two-Sided Network Effects," 1496.

15. Parker and Van Alstyne also note that this has effects on how monopoly is treated: "Platform intermediaries operating in two-sided markets seek to profit by transferring surplus from seller to consumer. Growth on one side of a matched market then induces growth on the other, creating exploitable surplus. This effect can prove unusually vexing to antitrust regulators because producer and consumer surplus can move together" (1495). That platform intermediaries tend toward monopoly is another trend noted in the literature on two- or multisided markets. Uber is one such intermediary, connecting willing drivers to people in need of a lift, but the model of commercial broadcast television is also an intermediary, connecting viewers' eyeballs to advertisers willing to pay for them.

16. Eisenmann, Parker, and Van Alstyne, "Strategies for Two-Sided Markets," trans. as "Tsū saido purattofōmu senryaku" [Two-sided platform strategies].

17. Eisenmann, Parker, and Van Alstyne, "Strategies for Two-Sided Markets," 94. See also Schmidt and Rosenberg, *How Google Works*, 79.

18. Eisenmann, Parker, and Van Alstyne, 96.

19. Eisenmann, Parker, and Van Alstyne, 94.

20. Evans and Schmalensee, *Matchmakers*, 20, emphasis mine.

21. Schmidt and Rosenberg, *How Google Works*, 79.

22. Interview with Kadokawa Tsuguhiko, Tokyo, June 26, 2013.

23. Negoro and Ajiro, "An Outlook of Platform Theory Research in Business Studies," 5. A Japanese working paper of this same article was published a year earlier as "Keieigaku ni okeru purattofomu ron no keifu to kongō no tenbō". [A genealogy of and future outlook on platform theory within management studies].

24. Negoro and Ajiro, "An Outlook of Platform Theory Research in Business Studies," 5.

25. Evans and Schmalensee, *Matchmakers*, 20.

26. Rysman, "The Economics of Two-Sided Markets," 127.

27. Rysman, 127.

28. Schmalensee, "Jeffrey Rohlfs' 1974 Model of Facebook," 301.

29. Rysman, "The Economics of Two-Sided Markets," 125.

30. Shapiro and Varian, *Information Rules*, 322.

31. Rohlfs, "A Theory of Interdependent Demand for a Communications Service," 16.

32. For an overview of this literature in relation to the VHS–Beta wars, see Evans and Schmalensee, *Matchmakers*, 22–30.

33. Katz and Shapiro, "Technology Adoption in the Presence of Network Externalities," 822.

34. Katz and Shapiro, "Network Externalities, Competition, and Compatibility," 424.

35. Rochet and Tirole, "Platform Competition in Two-Sided Markets," 990.

36. Rochet and Tirole, 1013–17.

37. *(Non)technological* would be another way to state the complex relation between technological system and nontechnological market or strategy.

38. Srnicek, *Platform Capitalism*, 45.

39. Eisenmann, Parker, and Van Alstyne, "Tsū saido purattofōmu senryaku," 68. It is unclear if this new preamble is written by the authors, the editors of the Japanese version of the *Harvard Business Review*, or the translator.

40. For platform capitalism I again refer to Srnicek but also Joss Hands; Trebor Scholtz's *Uberworked and Underpaid*; as well as the German tech writer Sasha Lobo, who likely coined the term in several 2014 contributions to *Der Spiegel*, including "Auf Dem Weg in die Dumpinghölle" [On the way to the dumping hole].

41. Negoro is also the coauthor of some of the most useful synthetic essays on the state of platform research, one of which is his and Ajiro Satoshi's "An Outlook of Platform Theory Research in Business Studies," around which I structure the division between three platform types.

42. Happily for readers who do not read Japanese, many of the writers discussed in this section have also published articles in English, some of which restate or translate in full arguments they make in the major works discussed here. I include references to these English sources when available.

43. An English version of a very similar argument and structure appears as the "Network Oriented Industrial Structures" chapter of Deguchi's *Economics as an Agent-Based Complex System*.

44. If this layered structure is reminiscent of the contents-platform structure, or hardware-software, it should be; Deguchi's subsequent work investigates contents.

45. Deguchi, "Nettowaku no rieki to sangyō kōzō," 42.

46. Deguchi, 42. For a recent account of the platform in relation to the stack, see Bratton, *The Stack*.

47. Deguchi, "Nettowaku no rieki to sangyō kōzō," 45. Of course, in the current era of the increasing deregulation of utilities this distinction no longer holds, and in fact utilities like electricity and gas are adopting more of a "platform" model of operation.

48. The earliest journal article to mention network externalities appears in 1990; soon after in the same year, Katz and Shapiro's original article also appears in Japanese translation.

49. Deguchi, "Kontentsu sangyō no purattofōmu kōzō to chōtayōsei chijō" [The platform structure and extreme variety market of the contents industry], 9.

50. Imai, "Hajime ni" [Preface], 3.

51. Imai, 3, emphasis mine.

52. Kokuryō, "Purattofōmu bijinesu to wa?" [What is a platform business?], 4.

53. Kokuryō, "Purattofōmu bijinesu no torihiki chūkai kinō to 'ōpun-kei keiei'" [Platform businesses as facilitators of transactions and their role in encouraging 'open' management], 12. Note that whereas in the English title accompanying his article Kokuryō translates his key term as "facilitators of transactions," a more streamlined translation, I prefer to preserve the mediatory quality of the transaction type through my rather direct and admittedly clunky translation: "transaction intermediary-type platform businesses."

54. Kokuryō, *Ōpun nettowāku keiei* [Open network management], 77–79.

55. Kokuryō, 149. Note that the above list and the sentence that follows are direct quotes.

56. Trust is a central element of most things we now call platforms—Airbnb, Uber, and so forth all require immense amount of trust between the user and the platform, which mediates and thereby creates trust between third parties who interact. I stay in your house not because I trust you but because I trust in the power and effectivity of the platform that mediates our relationship.

57. Kokuryō, *Ōpun nettowāku keiei*, 150.

58. Tapscott, *The Digital Economy*, 56.

59. Interview with Kokuryō Jirō, Tokyo, June 9, 2016.

60. Interview with Kokuryō Jirō, Tokyo, June 9, 2016. This initial study is Art Warbelow and Jiro Kokurō, "AUCNET TV Auction Network System."

61. Negoro and Kimura, *Netto bijinesu no keiei senryaku* [Management strategies of net businesses], 41.

62. Negoro and Kimura, 42.

63. Negoro and Kimura, 45.

64. Negoro and Kimura, 102.

65. Some of the early work on platforms in the U.S. context also focuses on the regulatory implications of platform businesses. This comes out of the impact of the antitrust suits against Microsoft and others, and reconsiders the monopoly concept in light of the platform research that would seem to advocate for "hubs" or "standards" that these platforms provide. Evans's work is particularly significant in this regard, insofar as he develops a robust typology of platforms as mediators, even as he considers them from the angle of monopoly law. Evans, "The Antitrust Economics of Multi-sided Platform Markets." Evans's coauthor on several books, Schmalensee, was in turn Microsoft's "star witness" during its antitrust trial of the 1990s.

66. Moazed and Johnson, *Modern Monopolies*, 17.

67. While I do not have the space to untangle these connections here, I will point briefly to the participation of Michael A. Cusumano in the International Motor Vehicle Program; Cusumano subsequently was the academic advisor to and coauthor with Annabelle Gawer, who became one of the first scholars to put the term *platform* onto the business studies map, albeit in a more computer-related, technical context at first. Gawer and Cusumano, *Platform Leadership*.

68. Japan's tech hub in Shibuya was known as Bit Valley (a play on the character "bitter" that makes up the "Shibu" part of Shibuya), for instance, and China's Shenzhen region has been called the "next Silicon Valley."

69. Gramsci, "Americanism and Fordism." This is also arguably an extension and transformation of the existing role of the corporation in social life. Braverman already notes "the immense amount of social coordination" exerted "by virtue of the giant size and power of the corporations, whose internal planning becomes, in effect, a crude substitute for necessary social planning." Braverman, *Labor and Monopoly Capital*, 186. Hence a closer analysis of this change would have to situate it within a larger history of the social coordination by corporations.

70. For both journalistic and critical accounts of platform disruption, see Stone, *The Upstarts;* Scholtz, *Uberworked and Underpaid*.

71. Tiziana Terranova's pathbreaking article on the subject, "Free Labor," begins as an account of AOL chat room volunteers realizing that they were in fact providing value to the company. That AOL was a web model that provided some inspiration for i-mode is of interest here, and suggests that we can reread hers and subsequent media studies work on digital labor (such as Andrejevic, "Exploiting YouTube") in light of the new managerial models of the platform that I explicate in this chapter. We can also read this change as being part of a longer history of the "audience commodity"—as Dallas W. Smythe calls it in his classic essay on television ("On the Audience Commodity and Its Work")—wherein what was once located in one sphere of the social life (television media consumption as multisided market wherein broadcasters sold eyeballs to advertisers) becomes generalized to other realms of consumption and work via the platformization of the economy.

72. Gillespie's *Wired Shut* provides a legal and engineering account of the hardware integration of legal contracts. The argument here is that management theory provided part of the justification for the expansion of control from workers to users/consumers to take place. For more on EULAs, see Perzanowski and Schultz, *The End of Ownership*.

4. Docomo's i-mode

1. Natsuno, *The i-mode Wireless Ecosystem*, 61–62.

2. See, for instance, Pelle Snickars and Patrick Vonderau's edited volume, *Moving Data*, which considers the future much more than the past of media that informs the iPhone itself, reinforcing the Steve Jobsian narrative line that the former Apple CEO was the inventor of everything. Fred Vogelstein's *Dogfight* is a useful popular account of the battle between Android and iOS; it offers an insightful account of their development but also neglects to mention the global impact of i-mode. Walter Isaacson's *Steve Jobs* offers an enchanting account of Jobs's genius but dwells little on his inspirations. One of the best general accounts of the smartphone that does acknowledge the pioneering role of i-mode and Japanese cell phones more generally is Elizabeth Woyke's *The Smartphone*. Larissa Hjorth's work on mobile media in Asia, Goggin's work on global mobile media, and Ito, Okabe,

and Matsuda's work on i-mode are all inspirations for and crucial resources for the work done in this chapter.

3. Woyke, *The Smartphone*, Kindle Location 522.

4. Hjorth, *Mobile Media in the Asia-Pacific*, 37.

5. Personal communication with Larissa Hjorth, May 6, 2017.

6. Goggin, *Global Mobile Media*, 139.

7. Natsuno, *Keitai no mirai* [The future of the mobile phone], 63.

8. As such they are a kind of prototypical digital commodity. See Gopinath, *The Ringtone Dialectic*.

9. Michael and Yutaka, "I-mode Is Coming"; Luna, "Coming to America"; Yutaka, Kimihide, and Tardy, "Lessons from Japan"; Rohwer et al., "On Top of the World"; Kushner, "Japan's Wireless Wonder"; Larimer, "Internet a la i-mode." These are just some of the examples of i-mode hype within both trade press and the mainstream magazines.

10. Gawer and Cusumano, *Platform Leadership*, 214–29.

11. Here I acknowledge that my focus on the business system rather than the use of the i-mode system risks painting it into too successful a light and repeating the technonationalism around i-mode in Japan of the early twenty-first century that Misa Matsuda critiques. Matsuda, "Discourses on *Keitai* in Japan," 32.

12. For Natsuno's account of his conversation with Schmitt, see Enoki, *I-mōdo no mōjū tsukai* [Training the i-mode beast], 182. For the global market share of Android, see, for instance, Vincent, "99.6 Percent of New Smartphones Run Android or iOS."

13. Following my argument in chapter 1 that we must take the Japanese pluralized form of contents seriously, and at face value, I use the term *contents provider* in what follows, except in cases where I quote from an English source that has rendered it as *content provider*. I should add that these singular forms of *content* are usually the conventions of translation, since most Japanese sources of the term I have been able to verify render the term as *kontentsu purobaidā* (contents provider) or *kontentsu jigyōsha* (contents businesses). This is especially true of the Japanese architects of i-mode, Natsuno Takeshi, Matsunaga Mari, and Enoki Kei'ichi, who uniformly use *kontentsu purobaidā* or *kontentsu jigyōsha* in Japanese.

14. Srnicek discusses the emergence of data as the most important resource for firms. Srnicek, *Platform Capitalism*, chapter 2.

15. Matsuda, "Discourses of *Keitai* in Japan," 32.

16. Fulford, "Docomo Phone Home." This figure is partly accounted for by Docomo holding the lion's share of the mobile market in toto.

17. Bradley and Sandoval, "NTT Docomo," n.p.

18. This leads Annabelle Gawer and Michael A. Cusumano to the hasty judgment of Docomo as a platform leader in Japan but a "platform wannabe" in the rest of the world. Gawer and Cusumano, *Platform Leadership*, 189.

19. Bradley and Sandoval, "NTT Docomo," n.p.

20. Natsuno, *The i-mode Wireless Ecosystem*, 133; Steinbock, *Wireless Horizon*, 163; Whittaker, *The Cyberspace Handbook*, 46.

21. Dornan, "I-mode Thinks Global Acts Local?," 24.

22. Kawakami and Kurita dialogue in Kawakami, *Gēmā wa motto keieisha wo mezasu beki!* [Gamers should aim to be managers!], 390.

23. Woyke, *The Smartphone*, Kindle Location 66. That said, the iPhone had a huge impact in the glass screen design and interface design that continues to inform the design grammar and the gestural interface of smartphones. Greenfield, *Radical Technologies*, 15.

24. Jin, *Smartland Korea*, 23.

25. Matsunaga Mari's *The Birth of i-mode* offers the best overview of the project.

26. Natsuno, *I-mode Strategy*, 125.

27. Natsuno's main advice to Japanese businesses based on his i-mode experience is to shift from thinking in terms of "making things" *(monozukuri)* to "making mechanisms" *(shikakezukuri)*. Natsuno, *Naze daikigyō ga totsuzen tsuburerunoka* [Why do massive companies suddenly go bust?], 80–81.

28. Kawakami, *Gēmā wa motto keieisha wo mezasu beki!*, 393.

29. Coates and Holroyd, *Japan and the Internet Revolution*, 75.

30. Kawakami, *Gēmā wa motto keieisha wo mezasu beki!*, 392. Some also claim that i-mode adds nothing new to DialQ2. See Misa Matsuda's comment on Kimura in "Discourses on *Keitai* in Japan," 39n13.

31. Evans, Hagiu, and Schmalensee, *Invisible Engines*, 294–95. According to Ken Coates and Carin Holroyd, "a typical i-mode user spends about 400 yen (US$3.25) a month on content subscriptions and 2,000 yen (US$17) on downloading content." Coates and Holroyd, *Japan and the Internet Revolution*, 78.

32. Natsuno quoted in Lynch and Clark, "Is Plain Old Cellular Enough for Wireless Data?," 29. In the same Lynch and Clark article, Natsuno is quoted as saying that "DoCoMo sets four criteria for what constitutes good content: "It needs to be updated more than once a day; It should be deep rather than shallow in length; It must be addictive; It must allow the user to hear or see the benefit."

33. Natsuno, *The i-mode Wireless Ecosystem*, 4.

34. Natsuno, xiv.

35. Natsuno, *I-mode Strategy*, 61–62; for the Japanese version of this English translation, see *I-mōdo sutorateji*, 118–19.

36. See James E. Moore, "Predators and Prey"; see also the book he wrote based on this article, *The Death of Competition*.

37. See Tiwana's *Platform Ecosystems* for a classic contemporary business account of platforms in relation to ecosystems.

38. Natsuno, *Keitai no mirai*, 200.

39. Natsuno, *The i-mode Wireless Ecosystem*, 19–20.

40. Evans, Hagiu, and Schmalensee, *Invisible Engines*, 256.

41. Interview with Natsuno Takeshi, Tokyo, January 6, 2016.

42. Interview with Kokuryō Jirō, Tokyo, June 9, 2016.

43. Natsuno, *1 chō yen wo kaseida otoko no shigotojutsu* [The work techniques of the man who made one billion yen], 160.

44. McKenna, *The World's Newest Profession,* 189. As McKenna documents, Boston Consulting Group, another of the major U.S. consultancies, had established a Tokyo office five years prior, in 1966.

45. Matsunaga, *The Birth of i-mode,* 31.

46. Matsunaga, 48.

47. Matsunaga, 39.

48. Enoki, *I-mōdo no mōjū tsukai,* 12.

49. Enoki, 12.

50. Enoki, 15.

51. Enoki, 33.

52. Matsunaga, *The Birth of i-mode,* 50–51.

53. Matsunaga, 46–47.

54. Matsunaga, 46, 49.

55. Schoenberger, "Queen of Mobile." See also Nanba's account of her creation of DeNA, *Fukakō keiei* [Ungainly management].

56. Obara, *IT bujinesu no genri* [The principles of the IT business], 26.

57. McKenna, *The World's Newest Profession,* 53. McKenna describes the concept of translators within Aitken's work as "those individuals who transfer a critical piece of information from one network to another" (277n11), referring the reader to Aitken, *The Continuous Wave.*

58. Kiechel, *The Lords of Strategy,* 191–92.

59. Evans and Schmalensee, *Matchmakers.*

60. Matsunaga herself notes that everyone insists on calling her the mother of i-mode, and acknowledges her discomfort at the phrase and her preference for the "developer of i-mode." Matsunaga, *The Birth of i-mode,* 7–8.

61. Quoted in Matsunaga, 78.

62. Matsunaga, 77.

63. Matsunaga, 85.

64. Quoted in Kridel, "I-opener," 28; also cited in Gawer and Cusumano, *Platform Leadership,* 224. See also Enoki's elaboration of this in *I-mōdo no mōjū tsukai,* 89–99, where he suggests the convenience store conception of i-mode as a major factor in its success.

65. Matsunaga, *The Birth of i-mode,* 51.

66. Matsunaga, 97.

67. Natsuno, *I-mode Strategy,* 63.

68. Srnicek, *Platform Capitalism,* 46.

69. Gawer and Cusumano, *Platform Leadership,* 225.

70. Ratliff, "NTT Docomo and Its i-mode Success," 60.

71. Kawakami and Kurita, *Gēmā wa motto keieisha wo mezasu beki!,* 385.

72. As Benjamin Fulford puts it, "It's a terrible arrangement for 40,000 'unofficial' sites—those not blessed by DoCoMo. They are raising a ruckus about how closed the system is. The unofficial sites aren't named on the phones' menu screens. The fastest way to access them is to manually enter their Web addresses. These sites have almost no way to charge for their services." Fulford, "Docomo Phone Home," 73.

73. Natsuno, *I-mode Strategy*, 104.

74. "Google Brings Award-Winning Search Engine to Japanese I-mode Users."

75. Hara, "Japan's i-mode Service Pioneers Shift to Packets," 61.

76. Normile, "Coming to America," 70.

77. Zhen and Ni, *Smart Phone and Next Generation Mobile Computing*, 146–47.

78. Quoted in Kridel, "I-opener," 24; see also Lamarre, *The Anime Ecology*.

79. Lynch and Clark, "Is Plain Old Cellular Enough for Wireless Data?," 29.

80. Young, "Settling Up," 24.

81. Enoki, *I-mōdo no mōjū tsukai*, 170.

82. For a book that covers—critically but journalistically—the mergers and acquisitions fervor of the late 1990s, especially in relation to what would be known as the content industries, see Seave, Greenwald, and Knee, *The Curse of the Mogul*.

83. Enoki, *I-mōdo no mōjū tsukai*, 171.

84. Bradley and Sandoval, "NTT Docomo," n.p.

85. Helmond, "The Platformization of the Web."

86. On appli coming preinstalled and supplanting the menu-based navigation process, see Ishino, *Mobage Taun ga sugoi riyū* [The reason why Mobage Town is amazing], 73–75.

87. Ironically one of the critics of the appification process is i-mode developer Kurita. Kawakami, *Gēmā wa motto keieisha wo mezasu beki!*, 394.

88. Clark, "The NTT Docomo Success Story," 48.

89. Natsuno, *I-mode Strategy*, 61–62.

90. Wu, *The Master Switch*, 262.

91. Rose, "Keyword: Context." Randall Stross describes AOL as selling consumers "pre-Web information services." Stross, *Planet Google*, 24.

92. Michael and Mizukoshi describe email as i-mode's killer app ("I-mode Is Coming," 74); Wu describes email as "AOL's most important feature" (*The Attention Merchants*, 204).

93. Ratliff, "NTT Docomo and Its i-mode Success," 59.

94. For a useful overview of each service, consult IBM's Red Paper: Megler, "The Semi-Walled Garden," 3.

95. Wu, *The Master Switch*, 261–63.

96. Natsuno, *iPhone vs. Andoroido* [iPhone vs. Android], 81–85.

97. "Global Mobile OS Market Share in Sales to End Users from 1st Quarter 2009 to 2nd Quarter 2017."

98. Natsuno remarks that the iPhone is an extension of i-mode. Natsuno, *Naze daikigyō ga totsuzen tsuburerunoka*, 79.

99. Kawakami, *Suzuki-san ni mo wakaru netto no mirai,* 100.

100. For an extended consideration of the lockout chip and the legal battles around it, see Sheff's classic account, *Game Over,* 161–71.

101. Kawakami, *Suzuki-san ni mo wakaru netto no mirai,* 145–46. Somewhat ironically, if Docomo's i-mode was inspired by Nintendo, the latter was in turn inspired by Docomo's parent company, telecommunications giant NTT, which had pioneered the "Captain System" in the early 1980s. While the Captain System (something like France's Minitel) was considered something of a failure, it generated much excitement and seemed to point to a future of platform-mediated communication—a future that Nintendo very much wanted to be part of. In the late 1980s, Nintendo aspired to be NTT, imagining a "Family Computer Communications Network" using its extensive network of existing Famicom consoles, as Sheff documents in *Game Over,* 76–78.

102. Evans, Hagiu, and Schmalensee, *Invisible Engines,* 125.

103. Evans, Hagiu, and Schmalensee, 152–53.

104. O'Donnell, "Production Protection to Copy(right) Protection." See also O'Donnell, "The Nintendo Entertainment System and the 10NES Chip."

105. Obara, who worked on the i-mode project as a McKinsey employee, also notes that i-mode is the model for both Apple and Google, which analyzed and modeled their own smartphones on i-mode. Obara, *Za purattofōmu* [The platform], 156.

106. Matsunaga notes that "with i-mode . . . although information could be downloaded from the Internet, a special device prevented copyrighted information from being extracted from the mobile phones." Matsunaga, *The Birth of i-mode,* 113.

107. Interview with Natsuno Takeshi, Tokyo, January 6, 2016.

108. A fourth and final model for i-mode is the 1980s "Captain System"—an early proto-internet home-based system that was developed by NTT (or its public sector predecessor Denden Kōsha). Similar to France's more successful Minitel project, the Captain System was an early form of "new media," as it was then called, forever giving the term a connotation that tied it the 1980s. Enoki, *I-mōdo no mōjū tsukai,* 41.

109. Kunii, "Japan's Cutting Edge."

110. Maamria, "New Era for Japan," 84.

111. Natsuno, *Naze daikigyō ga totsuzen tsuburerunoka,* 76–82. On Japan's concern with making things and the decline in the manufacturing sector, see Oyama's consideration of that discourse in light of the creative industries. Oyama, "Japanese Creative Industries in Globalization," 322–23.

112. Gopinath, *The Ringtone Dialectic,* 4.

113. Chun, *Updating To Remain the Same,* 8. *Her* and the television series *Black Mirror* are only two of the slew of thought experiments that trace out the consequences of mobile-everywhere and always-on culture; popular books on the

psychology of addiction written in relation to smartphones such as Alter's *Irresistible* trace this out at a different level.

5. Platforms after i-mode

1. As in the previous chapter, I use i-mode as the stand-in for other systems here and in what follows, given that it had a 60 percent market share for much of the period under question, it was the dominant system, and it was often the trendsetter as well.

2. Ohashi, Kato, and Hjorth, "Digital Genealogies," 5.

3. Natsuno, *The i-mode Wireless Ecosystem*, 40. I quote here from the 2003 English translation; the Japanese text was originally published in 2002. Ken Coates and Carin Holroyd estimate that some official sites made upward of US$4 million per month on subscriptions to their services. Coates and Holroyd, *Japan and the Internet Revolution*, 78.

4. Manabe, "Going Mobile," 318.

5. Ishikawa, *Web 2.0 jidai no keitai sensō* [The mobile wars of Web 2.0], 76.

6. Kawakami, *Gēmā wa motto keieisha wo mezasu beki!*, 391–92.

7. The *New York Times*, for instance, had 3.6 million total paid subscriptions and 2.6 million digital subscriptions as of the end of 2017, signaling a significant change from 2013, when Kurita believed Japan was unique in paying for news. This also signals the mobile-assisted transformation to a paid model of news. See Ember, "New York Times Co. Subscription Revenue Surpassed $1 Billion in 2017"; "The Times in Numbers 2017."

8. Mobile Content Forum, *Keitai hakusho 2011*, 150, 151.

9. Mobile Content Forum, *Sumaho hakusho 2015*, 35.

10. That said, the percentage occupied by the feature phone market has decreased as smartphones become the main means of accessing contents; in 2013 feature phone contents accounted for 2.4 billion yen, with smartphone contents at 8.3 billion yen. Mobile Content Forum, 35.

11. Gopinath, *Ringtone Dialectic*, 3.

12. Obara, *Za purattofōmu*, 167.

13. Gopinath, *Ringtone Dialectic*, 181.

14. On the subject of mechanisms, I refer the reader back to the previous chapter and its discussion of Natsuno's emphasis on "making mechanisms" over "making things." Natsuno, *Naze daikigyō ga totsuzen tsuburerunoka*, 80–81.

15. Kunii, "Japan's Cutting Edge," 160.

16. Coates and Holroyd quote a 2001 *Economist* article that states the following: "Internet users expect things to be free, and are prepared to accept a certain degree of technological imperfection. Mobile users are accustomed to paying, but expect a far higher level of service and reliability in return." Standage, "The Internet, Untethered," 5. The latter half is an apt description for the shift toward paid reliability rather than freeness as the basis for accessing contents.

17. Technically, Cybird was founded as Paradise Web in 1994, but it rebranded as Cybird for the i-mode project and only seems to have begun its business ventures at that point. Stewart, "Why Do Cybirds Suddenly Appear?"

18. Natsuno, *The i-mode Wireless Ecosystem,* 54.

19. "Andy Nulman."

20. Okada, "Youth Culture and the Shaping of Japanese Mobile Media"; Manabe, "Going Mobile," 319.

21. Kawakami, *Gēmā wa motto keieisha wo mezasu beki!,* 385. As a Japanese tech magazine writes: "Cybird is the premier (and first mover) content provider for NTT's DoCoMo i-mode and other operator services. Cybird is also a trusted source from DoCoMo's point of view, which means other content providers wanting a prize spot on the small screen are well served by going through Cybird." "Five Hot Startups."

22. "Riding the Wave, Carefully."

23. Rifkin, *The Age of Access.*

24. Natsuno, *I-mode Strategy,* 35; Natsuno, *The i-mode Wireless Ecosystem,* 51–52.

25. Galapagos discourse was being used around 2006 but was put into wider popular circulation by the Nomura Research Institute, Japan's largest information technology consultancy, around 2007, in its publication *2015nen no Nihon* [Japan in 2015].

26. In discussions of the entrance of the smartphone into Japan, it bears noting that while smartphones appeared to be cutting edge—they had large screens and surfed all the internet, not just those sites formatted for cHTML-compatible phones—they also lagged behind Japan's feature phones in many respects. Payment systems, for instance, introduced back in 2004 took some time to be implemented on the newer smartphones; the latter did not have the television-ready functionality of the feature phones, and so on. Natsuno Takeshi—admittedly a biased source—suggests that Japan's feature phones had by 2003 implemented Flash and had the same level of application development that smartphones reached only in 2011. Natsuno, *iPhone vs. Andoroido,* 63.

27. Helmond, "The Platformization of the Web," 1.

28. Nieborg and Poell, "The Platformization of Cultural Production," 2.

29. Interview with Natsuno Takeshi, Tokyo, January 5, 2016.

30. Ishikawa, *Web 2.0 jidai no keitai sensō,* 15–16; Ishino, *Mobage Taun ga sugoi riyū,* 20.

31. Ishino, *Mobage Taun ga sugoi riyū,* 43.

32. Ishino, 45–46.

33. Ishino, 19–21.

34. Natsuno, *iPhone vs. Andoroido,* 127. Natsuno's critical turn in his accounts of i-mode generally coincides with his departure from Docomo, so this critique may be read in that light. While his early accounts are incredibly celebratory, later accounts begin to be more critical, particularly around the year 2006, after which

point Natsuno feels like i-mode lost its direction and its drive toward increasing services.

35. Nakayama, *The Third Wave of Japanese Games,* Kindle Location 56.

36. Sasaki Toshinao, *Niconico dōga ga mirai wo tsukuru* [Niconico video is creating the future], 220.

37. Nakayama, *The Third Wave of Japanese Games,* Kindle Location 440.

38. "Business Integration of Kadokawa-Dwango."

39. Ōtsuka, "Kigyō ni kanri sareru saiteki na posutomodan no tame no essei" [An essay for the postmodern optimization of control by companies]. This blog post was published in its entirety as the introduction to Ōtsuka, *Media mix ka suru nihon* [The media mixization of Japan], 18.

40. Ōtsuka, "Kigyō ni kanri sareru"; Ōtsuka, *Media mix ka suru nihon,* 17. See also Álvaro David Hernández Hernández's analysis of this merger and its aftermaths, which further engages with Ōtsuka's work, in "The Anime Industry, Networks of Participation, and Environments for the Management of Content in Japan."

41. Interview with Kawakami Nobuo, Tokyo, Dec. 1, 2015.

42. Kawakami, *Suzuki-san ni mo wakaru netto no mirai,* 110.

43. "Dwango Company History."

44. Sasaki Toshinao, *Niconico dōga ga mirai wo tsukuru,* 219.

45. Sasaki Toshinao, 222.

46. Sasaki Toshinao, 232–34.

47. "The First Global 100," 115. For an in-depth analysis of Netflix as global television, see Ramon Lobato's *Netflix Nations.* Google CEO Eric Schmitt famously called these companies the Gang of Four at the "D: All Things Digital" conference in 2011. See also Kadokawa Tsuguhiko, *Google, Apple ni makenai chosakken*; Simon, *The Age of the Platform.*

48. For accounts of the media mix, see Joo, Denison, and Furukawa, "Manga Movies Project"; Lamarre, *The Anime Ecology*; Steinberg, *Anime's Media Mix*; and Zahlten, *The End of Japanese Cinema.*

49. Wollen, "Godard and Counter-Cinema"; Galloway, *Gaming,* 107–26.

50. The significant exception here is in China, where a "forked" version of Android is used that does not lock users into the Google ecosystem.

51. Jameson, "Notes on Globalization as a Philosophical Issue," 58.

52. Jameson, 63

53. Jin, "The Construction of Platform Imperialism in the Globalization Era," 154. Jin elaborates the concept further in his monograph *Digital Platforms, Imperialism, and Political Culture.*

54. David Golumbia addresses the cultural homogenizing tendencies of the computer code's reliance on English as its programming language, arguing that this centrality of English threatens the "loss of cultural diversity and a coordinate loss of linguistic diversity, and the coordination between these two is complex." He also implicitly draws on the cultural imperialism hypothesis in the following

formulation: "Like Hollywood films and U.S. television, the computer exerts a powerful attractive pull toward English that is perceived by many non-English-speakers as 'modern opportunity.'" Golumbia, *The Cultural Logic of Computation,* 123, 122.

55. See Sakai, *Translation and Subjectivity.*

56. "Top Sites in Japan."

57. Kadokawa Dwango, "KADOKAWA DWANGO Financial Results for the Fiscal Year Ended March 31, 2017," 17.

58. Interview with Kawakami Nobuo, Tokyo, Dec. 1, 2015.

59. As I noted in the introduction, Barubora and Sayawaka's *Bokutachi no intānetto shi* addresses the mobile internet only in the concluding 20 pages of a 230-page book—despite otherwise carefully going over developments decade by decade. There they situated the "semi-closed services" of the feature phone as a prehistory of social networking services, including Niconico. Yet, despite this, Niconico is discussed early on in the book as a crucial part of their internet history. This contradiction is never addressed head on. Barubora and Sayawaka, *Bokutachi no intānetto shi,* 212.

60. Sasaki Toshinao provides a useful account of the development of Niconico. Sasaki Toshinao, *Niconico dōga ga mirai wo tsukuru.*

61. Condry, *The Soul of Anime,* 63. In *The Anime Ecology,* Lamarre extends this analysis of character as platform.

62. Daniel Johnson offers a rich consideration of the comments function. Johnson, "Polyphonic/Pseudo-synchronic."

63. Describing it as unique may be a misnomer at this point. As I noted above, Chinese websites AcFun and later Bilibili have deployed the comments function and in case of Bilibili the very interface of Niconico; so the very peculiarity of the Niconico interface is in the process of being generalized into a new form of viewing experience.

64. Sasaki Toshinao, *Niconico dōga ga mirai wo tsukuru,* 272.

65. One significant exception to this that I cannot address at length here is Twitch Plays, wherein a game is streamed to Twitch, and the platform aggregates viewers' comments as control inputs. Here we find a form of commentary as play, or commentary as contents manipulation. Thanks to David Leblanc for noting this important exception.

66. Of the massive literature on the topic, I note Mark Andrejevic's useful overview and expansion of the argument to include data-mining. Andrejevic, "Exploiting YouTube."

67. Galloway, *The Interface Effect,* 40.

68. Galloway, 42.

69. Galloway, 42.

70. Cho, "Pop Cosmopolitics and K-pop Video Culture," 244. Here I also take inspiration from Goggin and McLelland's "Internationalizing Internet Studies," as well as Neves and Sarkar, "Introduction."

71. Nozawa, "The Gross Face and Virtual Fame."

72. Daniel Johnson, "Polyphonic/Pseudo-synchronic," 309.

73. Sasaki Toshinao notes the distinct cultures of the mobile-phone-based web culture from that of the PC-based web experience (interview with Sasaki Toshinao, Tokyo, Nov. 10, 2015). The distinction between the two types of users and uses is also noted by Kawakami himself, describing the PC-based users as "net-dwellers" or "net natives," while those who use the internet occasionally or as a tool (to send email, and so forth) are the "net toolers." Kawakami, *Suzuki-san ni mo wakaru netto no mirai*, 20.

74. Coleman, *Hacker, Hoaxer, Whistleblower, Spy*, 41.

75. Angela Nagle offers a useful critique and historical contextualization of this space in relation to the rise of the ultra-right and in relation to previous political movements. Nagle, *Kill All Normies*.

76. Kitada, "Japan's Cynical Nationalism," 70.

77. Kitada, 77.

78. Luhmann, "What Is Communication?"

79. Kitada, "Japan's Cynical Nationalism," 80. The figuration of the web as a space of communication for communication's sake is shared by Kitada's colleague Azuma Hiroki, one of the most prominent media theorists of the early twenty-first century, as well as by Jodi Dean. See Azuma, *Gēmu-teki riarizumu no tanjō* [The birth of gameic realism], 143–52; Dean, *Blog Theory*, 102.

80. The account of Niconico's development here is based on Sasaki Toshinao, *Niconico dōga ga mirai wo tsukuru*, 234–69.

81. Hamano, *Ākitekucha no seitaikei* [Ecosystems of architecture], 211. Sasaki Toshinao also uses the term in *Niconico dōga ga mirai wo tsukuru*, 259.

82. Deleuze, *Cinema 1*, 5.

83. By regional I refer to the communities of participation; while Japan is the principal place from which the site is accessed, the site also offers Taiwan and the United States as other regions and languages through which to access the site. It is also increasingly making inroads into Europe where communities of fandom around anime-manga-games from Japan are particularly strong. The region here is defined by affective communities of fandom rather than geography per se. For an extended consideration of the intersection of the media mix and regionalism, see Lamarre, "Regional TV."

84. For an analysis of the Kagerou Project from the perspective of the vocaloid phenomenon, see Shiba, *Hatsune Miku wa naze sekia wo kaetanoka?* [Why did Hatsune Miku change the world?], 218–25.

85. Li, "The Interface Affect of a Contact Zone," 235–36.

86. Li, 235–36.

87. This seems to be indebted to Korea's earlier social media platform, Cyworld, though I do wonder whether Niconico was also an influence here. Thanks to Michelle Cho for this note about Cyworld. On AfreecaTV, see Nam and Kwon, "How Important Are Real-Time Communications in TV Watching?," 2.

88. Hamano, "Tsukuru ākitekuchā wo tsukuru" [Making architecture for making], 48.

89. Natsuno, *Naze daikigyō ga totsuzen tsuburerunoka,* 80–81.

90. Natsuno, *I-mode Strategy,* 111–12.

91. Tiziana Terranova's classic essay "Free Labor" remains the benchmark of this critique, taken up in relation to platform capitalism by many. See also Snricek, *Platform Capitalism.*

92. Kawakami argues that Japan is a leader in net-born culture, because of its good infrastructure and the large number of NEETs. Kawakami, *Suzuki-san ni mo wakaru netto no mirai,* 25.

93. Ching, "Globalizing the Regional, Regionalizing the Global," 237.

94. Ching, 237.

95. Tsing, "Supply Chains and the Human Condition."

Conclusion

1. Lazzarato, *Les revolutions du capitalisme,* 96.

2. Gawer and Cusumano, *Platform Leadership,* 189–244.

3. David B. Nieborg's insightful work on the blockbuster economy of game apps informs this passage. See Nieborg, "Crushing Candy" and "The Political Economy of Platformed Production and Circulation."

4. Lobato and Meese, *Geoblocking and Global Videoculture.* Lobato defines geoblocking in the following manner: "Geoblocking, a spatially-aware filtering technology that uses IP address databases to determine a user's location, has become a key mechanism for managing international video streaming traffic and maintaining separation of national media markets" (10).

5. Two very useful critiques of the area studies paradigm remain Harry Harootunian's *History's Disquiet,* which offers a sustained critique of East Asian studies in particular and area studies in general as complicit with the U.S. Cold War paradigm of knowledge formation, and Rey Chow's *The Age of the World Target,* which introduces the generative concept of targeting and self-referentiality within this paradigm.

6. Ching, "Globalizing the Regional, Regionalizing the Global," 237.

7. Ching, 237.

8. Ching, 237.

9. Ching, 243.

10. Ching, 255.

11. Ching, 257.

12. Choi, "Of the East Asian Cultural Sphere," 116.

13. Choi, 120, 118.

14. Choi, 122.

15. Choi, 129.

16. Lamarre, "Regional TV," 94.

17. Lamarre, 109.

18. Lamarre, 113.

19. DeBoer, *Coproducing Asia*, 2.

20. DeBoer, 10.

21. On geofencing in the context of streaming apps, see Burroughs and Rugg, "Extending the Broadcast."

22. Light, Burgess, and Duguay, "The Walkthrough Method," 4.

23. Burroughs and Rugg, "Extending the Broadcast," 377.

24. Jin and Yoon, "Reimagining Smartphones in a Local Mediascape," 511.

25. That said, this platformization of the chat app is also appearing more and more a part of the Facebook Messenger strategy, so this is by no means exclusive to Asia, albeit it seems to have developed there first.

26. DeBoer, *Coproducing Asia*, 9.

27. "LINE Conference Tokyo 2014."

28. "LINE Conference Tokyo 2017."

29. Shin and Ha, *Yabai LINE* [Dangerous LINE], 7.

30. "LINE Conference Tokyo 2017."

31. Even as late as 2014, the feature phone market was larger in Japan (54 percent) than the smartphone market (46 percent). Mobile Content Forum, *Sumaho hakusho 2015*, 112. In 2016 the smartphone share of the market was 82 percent, indicating that feature phone sales are still a substantial part of the market at 18 percent. "Summary of Consolidated Financial Results for the Three Months Ended March 31, 2017," 5.

32. Natsuno, *Atarimae no senryaku shikō* [Obvious strategic thinking], 80–81.

33. "Top Sites Ranking for All Categories in Korea, Republic Of."

34. Jin and Yoon, "Reimagining Smartphones in a Local Mediascape," 7.

35. Shin and Ha, *Yabai LINE*, 137–38.

36. For 2014, I draw from Jin, *Smartland Korea*, 95; for 2015, see "2015 Search Engine Market Share by Country."

37. For an account of Livedoor's impact on the Japanese tech scene, and its subsequent decline, see Oyama, "Japanese Creative Industries in Globalization."

38. Shin and Ha, *Yabai LINE*, 143–44. According to Shin and Ha's account, Naver Korea had internal company plans to release a chat app to compete with Kakao—something that was eventually released as NaverTalk in February 2011.

39. "LINE: Dossier," 9.

40. According to a 2015 dossier, Richardson, "Mapping Out the Chat App Landscape."

41. "Tencent Investor Kit."

42. "110 Amazing WeChat Statistics and Facts"; "Most Popular Asia-Based Mobile Messenger Apps"; "WeChat Daily Active Users Reached 570 Mln in Sep 2015."

43. Yano, *Pink Globalization*.

44. Meeker, "Internet Trends 2015."

45. Interview with Natsuno Takeshi, Tokyo, Jan. 6, 2016; interview with LINE's manager of the sticker business planning team, Watanabe Naotomo, Tokyo, Jan. 6, 2016.

46. Ohashi, Fumitoshi, and Hjorth, "Digital Genealogies," 4.

47. See https://www.linefriends.com/, accessed June 17, 2017. The website has since redescribed the LINE friends as a "Global Character Brand," perhaps in light of an increasing target of North America and Europe as a site for LINE Friends stores.

48. Prigg, "Could Brown Bear and Cony the Bunny Replace Emoji?"

49. Interview with Watanabe Naotomo, Tokyo, Jan. 5, 2016.

50. "Big character" is a felicitous term suggested to me by an anonymous man in the audience at a talk I gave at the Japanese Association for Contents History Studies, Tokyo, Nov. 28, 2015.

51. The 2013 figures are from Shin and Ha, *Yabai LINE*, 42; the 2015 figures are from an email exchange with Icho Saito, Global PR for LINE, Mar. 29, 2016.

52. "Summary of Consolidated Financial Results for the Three Months Ended March 31, 2017," 5.

53. Thanks to Michelle Cho for pointing out the BTS stickers.

54. This is given as one of the reasons for using KakaoTalk by one of Jin and Yoon's interviewees: "Above all, as a 24-year-old respondent noted, 'the huge popularity of KakaoTalk is not because of technological quality but just the first mover's advantage' (Interviewee 7). KakaoTalk, as the first mover, has succeeded in securing a large user base and thus dominating the local smartphone app market." Jin and Yoon, "Reimagining Smartphones in a Local Mediascape," 6.

55. "Characters," Condry suggests, "offer a way to think about anime as a 'generative platform.'" Condry, *The Soul of Anime*, 58.

56. Horwitz, "One Year after the Government Banned Its Chat App."

57. Lamarre, "Regional TV," 94.

58. An important counternarrative to the emphasis on Silicon Valley companies is the narrative about its peoples, and its diaspora, about which AnnaLee Saxenian's *The New Argonauts* provides a fantastically destabilizing alternative narrative.

59. Again, this book is not by any means the only one to do so, and other fellow travelers include Mailland and Driscoll's *Minitel* and many other contributions to the MIT Platform Studies series, Larissa Hjorth's work on mobile media in Asia, Tom Lamarre's *The Anime Ecology*, and, in a more popular vein, Tim Wu's *Attention Merchants*.

60. See, for instance, *The Economist*'s accelerating coverage of Chinese technology giants, such as "China's Tech Industry Is Catching Up with Silicon Valley." The focus on Chinese tech is refreshing, albeit familiar in rhetorical mode to that around i-mode in Japan and earlier moments of Japanese hardware development.

61. Ching, "Neo-regionalism and Neoliberal Asia."

62. Bilton, "The Management of the Creative Industries," 31.

Bibliography

Aitken, Hugh G. J. *The Continuous Wave: Technology and American Radio, 1900–1932*. Princeton, N.J.: Princeton University Press, 1985.

Allison, Anne. "The Cool Brand, Affective Activism and Japanese Youth." *Theory, Culture & Society* 26, nos. 2–3 (2017): 89–111.

Alter, Adam. *Irresistible: The Rise of Addictive Technology and the Business of Keeping Us Hooked*. New York: Penguin, 2017.

Altice, Nathan. *I Am Error*. Cambridge, Mass.: MIT Press, 2015.

Anand, Bharat. *The Content Trap: A Strategist's Guide to Digital Change*. New York: Random House, 2016.

Andreessen, Marc. "Analyzing the Facebook Platform, Three Weeks In." Pmarca blog, June 12, 2007. http://web.archive.org/web/20071021003047/http://blog.pmarca.com/2007/06/analyzing_the_f.html.

Andreessen, Marc. "The Three Kinds of Platforms You Meet on the Internet." Pmarca blog, Sep. 16, 2007. http://web.archive.org/web/20071018161644/http://blog.pmarca.com/2007/09/the-three-kinds.html.

Andrejevic, Mark. "Exploiting YouTube: Contradictions of User-Generated Labour." In *The YouTube Reader*, edited by Pelle Snickars and Patrick Vondera, 406–23. Stockholm: National Library of Sweden, 2009.

"Andy Nulman." *The Art Of*. Accessed June 26, 2017. https://www.theartof.com.

Apperley, Thomas, and Jussi Parikka. "Platform Studies' Epistemic Threshold." *Games and Culture* 13, no. 4 (2015): 349–69.

Apprich, Clemens. *Technotopia: A Media Genealogy of Net Cultures*. London: Rowman & Littlefield, 2017.

Arai Noriko, Fukuda Toshihiko, and Yamakawa Satoru. *Kontentsu māketingu: Monogatari-kei shōhin no shijō hōsoku wo sagaru* [Contents marketing: Searching for the laws of the narrative commodity market]. Tokyo: Dobunkan, 2004.

Ashraf, Cameran, and Luis Felipe Alvarez Leon. "The Logics and Territorialities of Geoblocking." In *Geoblocking and Global Videoculture,* edited by Ramón Lobato and James Meese, 42–53. Amsterdam: Institute of Network Cultures, 2016.

Azuma Hiroki. *Gēmu-teki riarizumu no tanjō* [The birth of gameic realism]. *Dōbutsukasuru posutomodan 2* [The animalizing postmodern 2]. Tokyo: Kōdansha, 2007.

Azuma Hiroki, ed. *Kontentsu no shisō: Manga anime raitonoberu* [The thought of contents: Manga, anime, and light novels]. Tokyo: Seidosha, 2007.

Baldwin, Carliss Y., and C. Jason Woodard. "The Architecture of Platforms: A Unified View." In *Platforms, Markets and Innovation,* edited by Annabelle Gawer, 19–44. Cheltenham: Edward Elgar, 2009.

Baldwin, Matt. "Did Shakespeare Produce Good 'Content'?" Mar. 21, 2017. http://insights.coastcommunications.co.uk/post/102e2z7/did-shakespeare-produce-good-content.

Barubora and Sayawaka. *Bokutachi no intānetto shi* [Our internet history]. Tokyo: Akishobo, 2017.

Baudrillard, Jean. *The Consumer Society: Myths and Structures.* Translated by J. P. Mayer. London: SAGE, 1998.

Baudrillard, Jean. *The System of Objects.* Translated by James Benedict. London: Verso, 1996.

Bilton, Chris. "The Management of the Creative Industries." In *Managing Media Work,* edited by Mark Deuze, 31–42. Thousand Oaks, Calif.: SAGE, 2011.

Bogost, Ian, and Nick Montfort. "Platform Studies: Frequently Questioned Answers." *Digital Arts and Culture,* Dec. 12–15, 2009. http://www.bogost.com.

Bogost, Ian, and Nick Montfort. *Racing the Beam: The Atari Video Computer System.* Cambridge, Mass.: MIT Press, 2009.

Boltanski, Luc, and Eve Chiapello. *The New Spirit of Capitalism.* London: Verso, 2005.

Bradley, Stephen P., and Matthew Sandoval. "NTT Docomo: The Future of the Wireless Internet?" 2000. Harvard Business School case study 9-701-013.

Bratton, Benjamin H. *The Stack: On Software and Sovereignty.* Cambridge, Mass.: MIT Press, 2015.

Braverman, Harry. *Labor and Monopoly Capital: The Degradation of Work in the Twentieth Century.* 1974. Reprint, New York: Monthly Review Press, 1998.

Bresnahan, Timothy F., and Shane Greenstein. "Technological Competition and the Structure of the Computer Industry." *Journal of Industrial Economics* 47, no. 1 (1999): 1–40.

Burroughs, Benjamin, and Adam Rugg. "Extending the Broadcast: Streaming Culture and the Problems of Digital Geographies." *Journal of Broadcasting and Electronic Media* 58, no. 3 (2014): 365–80.

Castells, Manuel. *The Rise of the Network Society.* The Information Age: Economy, Society and Culture 1. Oxford: Blackwell, 1996.

Chandler, Alfred. *Inventing the Electronic Century.* New York: Free Press, 2001.

"China's Tech Industry Is Catching Up with Silicon Valley." *Economist,* Feb. 16, 2018.

Ching, Leo. "Globalizing the Regional, Regionalizing the Global: Mass Culture and Asianism in the Age of Late Capital." *Public Culture* 12, no. 1 (2000): 233–57.

Ching, Leo. "Neo-regionalism and Neoliberal Asia." In *Routledge Handbook of New Media in Asia,* edited by Larissa Hjorth and Olivia Khoo, 39–52. Abington: Routledge, 2016.

Cho, Michelle. "Pop Cosmopolitics and K-pop Video Culture." In *Asian Video Cultures,* edited by Joshua Neves and Bhaskar Sarkar, 240–65. Durham, N.C.: Duke University Press, 2017.

Choi, JungBong. "Of the East Asian Cultural Sphere: Theorizing Cultural Regionalization." *China Review* 10, no. 2 (2010): 109–36.

Choo, Kukhee. "Nationalization 'Cool': Japan's Government's Global Policy towards the Content Industry." In *Popular Culture and the State in East and Southeast Asia,* edited by Nissim Otmazgin and Eyal Ben-Ari, 83–103. London: Routledge, 2011.

Chow, Rey. *The Age of the World Target: Self-Referentiality in War, Theory, and Comparative Work.* Durham, N.C.: Duke University Press, 2006.

Chun, Wendy Hui Kyong. *Programmed Visions: Software and Memory.* Cambridge, Mass.: MIT Press, 2011.

Chun, Wendy Hui Kyong. *Updating to Remain the Same: Habitual New Media.* Cambridge, Mass.: MIT Press, 2016.

Clark, Robert. "The NTT Docomo Success Story." *America's Network* 104, no. 4 (2000): 46–49.

Coates, Ken, and Carin Holroyd. *Japan and the Internet Revolution.* London: Palgrave Macmillan, 2003.

Coleman, Gabriella. *Hacker, Hoaxer, Whistleblower, Spy: The Many Faces of Anonymous.* London: Verso, 2014.

Condry, Ian. "Anime Creativity: Characters and Premises in the Quest for Cool Japan." *Theory, Culture & Society* 26, nos. 2–3 (2009): 139–63.

Condry, Ian. *The Soul of Anime: Collaborative Creativity and Japan's Media Success Story.* Durham, N.C.: Duke University Press, 2013.

Cowan, Deborah. *The Deadly Life of Logistics.* Minneapolis: University of Minnesota Press, 2014.

Dean, Jodi. *Blog Theory.* Malden, Mass.: Polity, 2010.

DeBoer, Stephanie. *Coproducing Asia: Locating Japanese-Chinese Regional Film and Media.* Minneapolis: University of Minnesota Press, 2014.

Deguchi Hiroshi. *Economics as an Agent-Based Complex System: Towards Agent-Based Social Systems Sciences.* Tokyo: Springer-Verlag, 2004.

Deguchi Hiroshi. "Kontentsu sangyō no purattofōmu kōzō to chōtayōsei chijō" [The platform structure and extreme variety market of the contents industry]. In *Kontentsu sangyōron* [On the contents industry], edited by Deguchi Hiroshi, Tanaka Hideyuki, and Koyama Yūsuke, 3–40. Tokyo: Tokyo Daigaku Shuppan, 2009.

Deguchi Hiroshi. "Nettowaku no rieki to sangyō kōzō" [Network merits and industrial structure]. *Journal of the Japan Society for Management Information* 1, no. 2 (1993): 41–61.

Deguchi Hiroshi, Tanaka Hideyuki, and Koyama Yūsuke, eds. *Kontentsu sangyōron* [On the contents industry]. Tokyo: Tokyo Daigaku Shuppan, 2009.

Dejitaru Kontentsu Kyōkai. *Dejitaru kontentsu hakusho 2013* [Digital contents white paper 2013]. Tokyo: Dejitaru Kontentsu Kyōkai, 2013.

Deleuze, Gilles. *Cinema 1: The Movement Image.* Translated by Hugh Tomlinson and Barbara Habberjam. Minneapolis: University of Minnesota Press, 1986.

Deleuze, Gilles. *Foucault.* Translated by Sean Hand. Minneapolis: University of Minnesota Press, 1988.

Dentsū Sōken. *Jōhō media hakusho 1996* [Informational media white paper 1996]. Tokyo: Dentsū Sōken, 1996.

Dentsū Sōken. *Jōhō media hakusho 1998* [Informational media white paper 1998]. Tokyo: Dentsū Sōken, 1998.

Dentsū Kenkyū, ed. *Jōhō media hakusho 2008* [Informational media white paper 2008]. Tokyo: Diamond, 2008.

Dentsū Kenkyū, ed. *Jōhō media hakusho 2010* [Informational media white paper 2010]. Tokyo: Diamond, 2010.

Dentsū Kenkyū, ed. *Jōhō media hakusho 2011* [Informational media white paper 2011]. Tokyo: Diamond, 2011.

Dornan, Andy. "I-mode Thinks Global Acts Local?" *Network Magazine* 17, no. 4 (2002).

"Dwango Company History." Accessed June 17, 2017. http://dwango.co.jp/english/history/.

"Dwango wa ōhaba zōshū zōeki, Kadokawa wa akagi" [Dwango expands income and profit, Kadokawa is in the red]. *IT Media News,* Nov. 13, 2014. http://www.itmedia.co.jp/news/articles/1411/13/news136.html.

Eisenmann, Thomas R., Geoffrey Parker, and Marshall W. Van Alstyne. "Strategies for Two-Sided Markets." *Harvard Business Review* 84, no. 10 (2006): 92–101. Translated as "Tsū saido purattofōmu senryaku" [Two-sided platform strategies], *Harvard Business Review* (Japanese edition) 32, no. 6 (2007): 68–81.

Ember, Sydney. "New York Times Co. Subscription Revenue Surpassed $1 Billion in 2017." *New York Times,* Feb. 8, 2018.

Enoki Kei'ichi. *I-mōdo no mōjū tsukai* [Training the i-mode beast]. Tokyo: Kodansha, 2015.

Enterbrain. *Sekai no entame gyōkai chizu 2013 nenban* [World entertainment industry map, 2013 edition]. Tokyo: Kadokawa, 2013.

Evans, David S. "The Antitrust Economics of Multi-sided Platform Markets." *Yale Journal on Regulation* 20, no. 2 (2003): 324–81.

Evans, David S., Andrei Hagiu, and Richard Schmalensee. *Invisible Engines: How Software Platforms Drive Innovation and Transform Industries.* Cambridge, Mass.: MIT Press, 2006.

Evans, David S., and Richard Schmalensee. *Matchmakers: The New Economics of Multisided Platforms.* Boston: Harvard Business Review Press, 2016.

Ferguson, Tim W. "Business World: Out of FCC, Sikes Says Telecom Signals Still Strong." *Wall Street Journal,* Apr. 6, 1993, A15.

"The First Global 100: Reed Hastings." *Wired UK,* Sep. 2016, 115.

"Five Hot Startups." *J@pan Inc,* Sep. 2000. http://www.japaninc.com.

Flint, Jerry, and Bob Tomarkin. "Wait 'til Next Year." *Forbes,* Oct. 15, 1979, 51–55.

Foucault, Michel. *The Order of Things: An Archaeology of the Human Sciences.* Translated by A. M. Sheridan Smith. New York: Vintage Books, 1994.

Frosh, Paul. *The Image Factory: Consumer Culture, Photography and the Visual Content Industry.* London: Berg, 2003.

Fujioka Wakao. *Fujioka Wakao zen purodūsu.* Vol. 2, *Mōretsu kara biūtifuru e* [Fujioka Wakao's productions. Vol. 2, From intense to beautiful]. Kyoto: PHP Kenkyūjo, 1988.

Fujioka Wakao. *Sayōnara, taishū: Kansei jidai o dō yomu ka* [Goodbye, masses: How to read the age of sensibility]. Reprint, Kyoto: PHP Bunko, 1987.

Fukuda Toshihiko. *Monogatari māketingu* [Narrative marketing]. Tokyo: Takeuchi Shoten, 1990.

Fukutomi Tadakazu. "Kontentsu to wa nanika?" [What is contents?]. In *Kontentsugaku* [Contents studies], edited by Hasegawa Fumio and Fukutomi Tadakazu, 2–17. Kyoto: Sekai shisōsha, 2007.

Fulford, Benjamin. "Docomo Phone Home." *Forbes,* May 14, 2001, 73.

Furuhata, Yuriko. "Architecture as Atmospheric Media: Tange Lab and Cybernetics." In *Media Theory in Japan,* edited by Marc Steinberg and Alexander Zahlten, 52–79. Durham, N.C.: Duke University Press, 2017.

Galbraith, Patrick, and Jason Karlin, eds. *Idols and Celebrity in Japanese Media Culture.* New York: Palgrave Macmillan, 2012.

Galloway, Alexander. *Gaming: Essays on Algorithmic Culture.* Minneapolis: University of Minnesota Press, 2006.

Galloway, Alexander. *The Interface Effect.* Malden, Mass.: Polity, 2012.

Ganti, Tejaswini. *Producing Bollywood: Inside the Contemporary Hindi Film Industry.* Durham, N.C.: Duke University Press, 2012.

Gardner, William O. "The 1970 Osaka Expo and/as Science Fiction." *Review of Japanese Culture and Society* (Dec. 2011): 26–43.

Gates, Bill. "Content Is King." Microsoft, Jan. 3, 1996. http://web.archive.org/web/20010126005200/http://www.microsoft.com/billgates/columns/1996essay/essay960103.asp.

Gawer, Annabelle. "Platform Dynamics and Strategies: From Products to Services." In *Platforms, Markets and Innovation,* edited by Annabelle Gawer, 45–63. Cheltenham: Edward Elgar, 2009.

Gawer, Annabelle, and Michael Cusumano. *Platform Leadership: How Intel, Microsoft, and Cisco Drive Industry Innovation.* Boston: Harvard Business Review Press, 2002.

Gillespie, Tarleton. "The Politics of 'Platforms.'" *New Media & Society* 12, no. 3 (2010): 347–64.

Gillespie, Tarleton. *Wired Shut: Copyright and the Shape of Digital Culture.* Cambridge, Mass.: MIT Press, 2007.

"Global Mobile OS Market Share in Sales to End Users from 1st Quarter 2009 to 2nd Quarter 2018." *Statista,* 2018. Accessed Aug. 20, 2018. https://www.statista.com.

Goggin, Gerald. *Global Mobile Media.* London: Routledge, 2011.

Goggin, Gerald, and Mark McLelland. "Internationalizing Internet Studies: Beyond Anglophone Paradigms." In *Internationalizing Internet Studies: Beyond Anglophone Paradigms,* edited by Gerald Goggin and Mark McLelland, 3–17. London: Routledge, 2009.

Golumbia, David. *The Cultural Logic of Computation.* Cambridge, Mass.: Harvard University Press, 2009.

"Google Brings Award-Winning Search Engine to Japanese I-mode Users." Google News, Feb. 27, 2001. http://googlepress.blogspot.ca.

Gopinath, Sumanth. *The Ringtone Dialectic: Economy and Cultural Form.* Cambridge, Mass.: MIT Press, 2013.

Gramsci, Antonio. "Americanism and Fordism." In *Selections from the Prison Notebooks,* translated and edited by Quintin Hoare and Geoffrey Nowell Smith, 279–322. New York: International Publishers, 1971.

Greenfield, Adam. *Radical Technologies.* London: Verso, 2017.

Grewal, David Singh. *Network Power: The Social Dynamics of Globalization.* New Haven, Conn.: Yale University Press, 2008.

Grove, Andy. "Need or Greed." *Multimedia and Videodisk Monitor Future Systems* 10, no. 9 (1992).

Guillory, John. "Genesis of the Media Concept." *Critical Inquiry* 36, no. 2 (2010): 321–62.

Hamano Satoshi. *Ākitekucha no seitaikei: Jō kankyo wa ikani sekkei sarete kita ka* [Ecosystems of architecture: How do information environments come to be planned?]. Tokyo: NTT Shuppan, 2008.

Hamano Satoshi. "Niconico dōga to Hatsune Miku" [Niconico Video and Hatsune Miku]. In *Niconico gakkai beta wo kenkyū shitemita Beta* [I tried to study Niconico beta conference], edited by Eto Kōichirō, 67–78. Tokyo: Kawade, 2012.

Hamano, Satoshi. "Tsukuru ākitekuchā wo tsukuru" [Making architecture for making]. In *Niconico gakkai beta wo kenkyū shitemita Beta* [I tried to study Niconico beta conference], edited by Eto Kōichirō, 46–66. Tokyo: Kawade, 2012.

Hands, Joss. "Introduction: Politics, Power and 'Platformativity.'" *Culture Machine* 14 (2013): 1–9.

Hara Yoshiko. "Japan's i-mode Service Pioneers Shift to Packets." *Electronic Engineering Times,* Aug. 21, 2000, 61.

Harootunian, Harry. *History's Disquiet: Modernity, Cultural Practice, and the Question of Everyday Life.* New York: Columbia University Press, 2002.

Hatakeyama Kenji and Kubo Masakazu. *Odoru kotentsu bijinesu no mirai* [Future of chaotic entertainment content biz]. Tokyo: Shōgakkan, 2005.

Hayase Goto. "The Current Condition and Framework of IPOs on Junior Markets in Japan." *Securities Analysts Journal* (May 2015): 1–13. https://www.saa.or.jp.

Hayashi Kōichirō. *Nettowākingu no keieigaku* [Management studies of networking]. Tokyo: NTT, 1989.

Helmond, Anne. "The Platformization of the Web: Making Web Data Platform Ready." *Social Media + Society* 1, no. 2 (2015): 1–11.

Hernández, Álvaro David Hernández. "The Anime Industry, Networks of Participation, and Environments for the Management of Content in Japan." *Arts* 7, no. 42 (2018): 1–20.

Hjorth, Larissa. *Mobile Media in the Asia-Pacific: Gender and the Art of Being Mobile*. London: Routledge, 2008.

Hjorth, Larissa, and Michael Arnold. *Online@AsiaPacific: Mobile, Social and Locative Media in the Asia-Pacific*. London: Routledge, 2015.

Holt, Jennifer, and Alisa Perren. "Introduction: Does the World Really Need One More Field of Study?" In *Media Industries: History, Theory, and Method*, edited by Jennifer Holt and Alisa Perren, 1–16. Malden, Mass.: Wiley-Blackwell, 2009.

Holt, Jennifer, and Kevin Sanson. *Connected Viewing: Selling, Streaming, & Sharing Media in the Digital Era*. New York: Routledge, 2014.

Horn, Robert E. *Haipātekisuto jōhō seirigaku: Kōzōteki kontentsu seisaku no susume* [Studies of hypertextual information organization: Recommendations on the creation of structural contents]. Translated by Matsubara Mitsuharu. Tokyo: Nikkei BP Shuppan, 1995.

Horn, Robert E. *Mapping Hypertext: The Analysis, Organization, and Display of Knowledge for the Next Generation of On-Line Text and Graphics*. Lexington: Lexington Institute, 1989.

Horwitz, Josh. "One Year after the Government Banned Its Chat App, Line Is Still in China—Selling Lattes and Tote Bags." *Quartz*, Aug. 3, 2015. http://qz.com.

Hyland, K. 1998. "Exploring Corporate Rhetoric: Metadiscourse in the CEO's Letter." *Journal of Business Communication* 35, no. 2: 224–44.

Idei Nobuyuki, ed. *Shinka suru purattofōmu: Gūguru Appuru Amazon o koete* [Evolving platforms: Overcoming Google, Apple, Amazon]. Tokyo: Kadokawa, 2015.

Imai Kenichi. "Hajime ni" [Preface]. In "Purattofōmu bijinesu" [Platform business], edited by Imai Kenichi and Kokuryō Jirō, special issue, *InfoCom REVIEW* (Winter 1994): 3.

Imai Kenichi and Kokuryō Jirō, eds. "Purattofōmu bijinesu" [Platform business]. Special issue, *InfoCom REVIEW* (Winter 1994).

Inamasu Tatsuo. *"Posuto koseika" no jidai: Kōdo shōhi bunka no yukue* [The age of post individualization: The whereabouts of high consumer society]. Tokyo: Jiji Tsūshin, 1992.

Isaacson, Walter. *Steve Jobs*. New York: Simon & Schuster, 2011.

Ishikawa Tsutsumu. _Web 2.0 jidai no keitai sensō_ [The mobile wars of Web 2.0]. Tokyo: Kadokawa, 2006.

Ishino Junya. _Mobage Taun ga sugoi riyū_ [The reason why Mobage Town is amazing]. Tokyo: Mainichi Communications, 2007.

Itō Hiroyuki and Hamano Satoshi. "Tsukuru akitekucha wo tsukuru" [Creating architecture to create]. In _Niconico gakkai Beta wo kenkyū shitemita_ [I tried to study Niconico beta conference], edited by Eto Kōichirō, 46–66. Tokyo: Kawade shobo, 2012.

Ito, Mizuko. Introduction to _Personal Portable Pedestrian_, edited by Mizuko Ito, Daisuke Okabe, and Misa Matsuda, 1–16. Cambridge, Mass.: MIT Press, 2005.

Ito, Mizuko. "Technologies of the Childhood Imagination: Yu-Gi-Oh!, Media Mixes, and Everyday Cultural Production." In _Structures of Participation in Digital Culture_, edited by Joe Karaganis, 88–111. New York: Social Science Research Council, 2007.

Ito, Mizuko, Daisuke Okabe, and Misa Matsuda, eds. _Personal Portable Pedestrian_. Cambridge, Mass.: MIT Press, 2005.

Ivy, Marilyn. "Critical Texts, Mass Artifacts: The Consumption of Knowledge in Postmodern Japan." In _Postmodernism and Japan_, edited by Masao Miyoshi and Harry D. Harootunian, 21–46. Durham, N.C.: Duke University Press, 1989.

Ivy, Marilyn. _Discourses of the Vanishing: Modernity, Phantasm, Japan_. Chicago: University of Chicago Press, 1995.

Ivy, Marilyn. "Formations of Mass Culture." In _Postwar Japan as History_, edited by Andrew Gordon, 239–58. Berkeley: University of California Press, 1993.

Jameson, Fredric. "Notes on Globalization as a Philosophical Issue." In _The Cultures of Globalization_, edited by Fredric Jameson and Masao Miyoshi, 54–77. Durham, N.C.: Duke University Press, 1998.

Jenkins, Henry. _Convergence Culture: Where Old and New Media Collide_. New York: New York University Press, 2006.

Jin, Dal Yong. "The Construction of Platform Imperialism in the Globalization Era." _tripleC_ 11, no. 1 (2013): 145–72.

Jin, Dal Yong. _Digital Platforms, Imperialism, and Political Culture_. New York: Routledge, 2015.

Jin, Dal Yong. _Smartland Korea: Mobile Communication, Culture, and Society_. Ann Arbor: University of Michigan Press, 2017.

Jin, Dal Yong, and Kyong Yoon. "Reimagining Smartphones in a Local Mediascape: A Cultural Analysis of Young KakaoTalk Users in Korea." _Convergence: The International Journal of Research into New Media Technologies_ 2, no. 5 (2016): 1–14.

Johnson, Chalmers. "Economic Crisis in East Asia: The Clash of Capitalisms." In _Financial Liberalization and the Asian Crisis_, edited by Ha-Joon Chang, Gabriel Palma, and D. Hugh Whittaker, 8–20. Basingstoke: Palgrave Macmillan, 2001.

Johnson, Chalmers. _MITI and the Japanese Miracle: The Growth of Industrial Policy, 1925–1975_. Stanford: Stanford University Press, 1982.

Johnson, Daniel. "Polyphonic/Pseudo-synchronic: Animated Writing in the Comment Feed of Nicovideo." *Japan Studies* 33, no. 3 (2013): 297–313.

Johnson, Derek, Derek Kompare, and Avi Santo, eds. *Making Media Work: Cultures of Management in the Entertainment Industries.* New York: New York University Press, 2014.

Joo, Woojeong, and Rayna Denison, with Hiroko Furukawa. "Manga Movies Project Report 1: Transmedia Japanese Franchising." 2013. mangamoviesproject.com. Archived at https://www.academia.edu/3693690/Manga_Movies_Project_Report_1_-_Transmedia_Japanese_Franchising.

"Kadokawa-Dowango keiei tōgō" [The business integration of Kadokawa-Dwango]. *Huffington Post Japan,* May 14, 2014. www.huffingtonpost.jp/2014/05/14/kadokawa-dwango-kawakami_n_5321576.html.

Kadokawa Dwango. "KADOKAWA DWANGO Financial Results for the Fiscal Year Ended March 31, 2017." May 11, 2017. https://info.kadokawadwango.co.jp/english/ir/pdf_kd/financial/20170511.pdf.

"Kadokawa Kaichō, keiei tōgō de 'Hi no maru purattofōmu tsukuru'" [Chairman Kadokawa, the merger will "create a platform of the rising sun"]. *Shinbunka,* May 5, 2014. http://www.shinbunka.co.jp/news2014/05/140514-06.htm.

Kadokawa Shoten. *Corporate Brochure.* Tokyo: Kadowaka Shoten, 2004.

Kadokawa Tsuguhiko. "Atarashii shuppan wo motomete" [In pursuit of new publishing]. In *Kadokawa Shoten sōritsu 40 shūnen* [Kadokawa Books 40 years from its founding], 2–8. Tokyo: Kadokawa Shoten, 1984.

Kadokawa Tsuguhiko. *Google, Apple ni makenai chōsakuken* [Copyright laws that can't be defeated by Google and Apple]. Tokyo: E-pub shinsho, 2013.

Kadokawa Tsuguhiko and Zenji Katagata. *Kuraudo Jidai to "Kūru Kakumei"* [The cloud era and the "cool revolution"]. Tokyo: Kadokawa Shoten, 2010.

Katz, Michael L., and Carl Shapiro. "Network Externalities, Competition, and Compatibility." *American Economic Review* 75, no. 3 (1985): 424–40.

Katz, Michael L., and Carl Shapiro. "Technology Adoption in the Presence of Network Externalities." *Journal of Political Economy* 94, no. 4 (1986): 822–41.

Kawakami Nobuo. *Gēmā wa motto keiesha wo mezasu beki!* [Gamers should aim to be managers!]. Tokyo: Kadokawa, 2015.

Kawakami Nobuo. *Suzuki-san ni mo wakaru netto no mirai* [The future of the internet that even Mr. Suzuki can understand]. Tokyo: Iwanami shinsho, 2015.

Kenney, Martin, and Richard Florida. *Beyond Mass Production: The Japanese System and Its Transfer to the U.S.* Oxford: Oxford University Press, 1993.

Kenney, Martin, and John Zysman. "The Rise of the Platform Economy." *Issues in Science and Technology* 32, no. 3 (2016): 61–69.

Kiechel, Walter. *The Lords of Strategy: The Secret Intellectual History of the New Corporate World.* Boston: Harvard Business Review Press, 2010.

Kitada, Akihiro. "Japan's Cynical Nationalism." In *Fandom Unbound: Otaku Culture in a Connected World,* edited by Mizuko Ito, Daisuke Okabe, and Izumi Tsuji, 68–84. New Haven, Conn.: Yale University Press, 2012.

Kittler, Friedrich A. *Gramophone, Film, Typewriter.* Translated by Geoffrey Winthrop-Young and Michael Wutz. Stanford: Stanford University Press, 1999.

"Kīwādo de kiku '95 jōhō tsūshin sangyō: Kontentsu" [Understanding the 1995 information communication industries by keyword: Contents]. *Nikkei shinbun,* Jan. 1, 1995, 6.

Kohara Hiroshi. *Nihon māketingu shi: Gendai ryūtsū no shiteki kōzu* [Japanese marketing history: The historical composition of contemporary distribution]. Tokyo: Chūō Keizaisha, 1994.

Kojima Tsuneharu. "Komāsharu to māketingu" [Commercials and marketing]. In *CM 25nen-shi* [A 25-year history of TV commercials], edited by Katō Shūsaku, 47–112. Tokyo: Kodansha, 1978.

Kokuryō Jirō. *Ōpun nettowāku keiei: Kigyō senryaku no shinchōryū* [Open network management: New trends in business strategy]. Tokyo: Nihon keizai shinbunsha, 1995.

Kokuryō Jirō. "Purattofōmu bijinesu no torihiki chūkai kinō to 'ōpun-kei keiei'" [Platform businesses as facilitators of transactions and their role in encouraging 'open' management]. In "Purattofōmu bijinesu" [Platform business], edited by Imai Kenichi and Kokuryō Jirō, special issue, *InfoCom REVIEW* (Winter 1994): 12–20.

Kokuryō Jirō. "Purattofōmu bijinesu to wa?" [What is a platform business?]. In "Purattofōmu bijinesu" [Platform business], edited by Imai Kenichi and Kokuryō Jirō, special issue, *InfoCom REVIEW* (Winter 1994): 4.

Kokuryō Jirō. "Purattofōmu ga sekai wo kaeru" [Platforms are changing the world]. In *Sōhatsu keiei no purattofōmu: Kyōdō no jōhō kiban zukuri* [Platforms for emergent management: Building the information basis for cooperative work], edited by Kokuryō Jirō and the Platform Design Lab, 1–12. Tokyo: Nihon keizai shinbunsha, 2011.

Kokuryō Jirō and the Platform Design Lab, eds. *Sōhatsu keiei no purattofōmu: Kyōdō no jōhō kiban zukuri* [Platforms for emergent management: Building the information basis for cooperative work]. Tokyo: Nihon keizai shinbunsha, 2011.

Kridel, Tim. "I-opener." *Wireless Review,* Oct. 1, 2000, 22–28.

Kunii, Irene M. "Japan's Cutting Edge." *Bloomberg,* May 21, 2000.

Kushner, Dave. "Japan's Wireless Wonder." *Rolling Stone,* no. 854 (Nov. 23, 2000): 7–8.

"Kyō no kotoba: Kontentsu" [Word of the day: Contents]. *Nikkei shinbun,* Apr. 2, 1995, 3.

Lamarre, Thomas. "Anime." In *The Japanese Cinema Book,* edited by Hideaki Fujiki and Alstair Reynolds. London: British Film Institute, forthcoming.

Lamarre, Thomas. *The Anime Ecology: A Genealogy of Television, Animation, and Game Media.* Minneapolis: University of Minnesota Press, 2018.

Lamarre, Thomas. *The Anime Machine.* Minneapolis: University of Minnesota Press, 2009.

Lamarre, Thomas. "Platformativity: Media Studies, Area Studies." *Asiascape: Digital Asia* 4, no. 3 (2017): 285–305.

Lamarre, Thomas. "Regional TV: Affective Media Geographies." *Asiascape: Digital Asia* 2, nos. 1–2 (2015): 93–126.

Larimer, Tim. "Internet a la i-mode." *Time* 157, no. 9 (2001): 38–40.

Latour, Bruno. *We Have Never Been Modern*. Translated by Catherine Porter. Cambridge, Mass.: Harvard University Press, 1991.

Lazzarato, Maurizio. *Les revolutions du capitalisme*. Paris: Les Empecheurs de Penser en Rond, 2004.

Ledonvirta, Vili, and Edward Castronova. *Virtual Economies: Design and Analysis*. Cambridge, Mass.: MIT Press, 2014.

Levinson, Andrew. "Who's Who on the Information Superhighway?" *Online* 19, no. 3 (1995): 98.

Li, Jinying. "The Interface Affect of a Contact Zone: *Danmaku* on Video-Streaming Platforms." *Asiascape: Digital Asia* 4, no. 3 (2017): 233–56.

Light, Ben, Jean Burgess, and Stefanie Duguay. "The Walkthrough Method: An Approach to the Study of Apps." *New Media & Society* 20, no. 3 (2016): 1–20.

"LINE: Dossier." *Statista*, 2017. Accessed June 17, 2017. https://www.statista.com/topics/1999/line/.

"LINE Conference Tokyo 2014." Accessed June 17, 2017. http://www.ustream.tv/recorded/53744592.

"LINE Conference Tokyo 2017." Accessed June 17, 2017. http://www.ustream.tv/recorded/104824368.

Liu, Alan. *The Laws of Cool: Knowledge and the Culture of Information*. Chicago: University of Chicago Press, 2004.

Liu, Alan. "Transcendental Data: Toward a Cultural History and Aesthetics of the New Encoded Discourse." *Critical Inquiry* 31, no. 1 (2004): 49–84.

Lobato, Ramon. *Netflix Nations: The Geography of Digital Distribution*. New York: New York University Press, 2019.

Lobato, Ramon, and James Meese, eds. *Geoblocking and Global Videoculture*. Amsterdam: Institute of Network Cultures, 2016.

Lobo, Sasha. "Auf Dem Weg in die Dumpinghölle" [On the way to the dumping hole]. *Der Spiegel*, Mar. 9, 2014.

Lovink, Geert. "Foreword." In *Technotopia: A Media Genealogy of Net Cultures*, by Clemens Apprich, xiii–xix. London: Rowman & Littlefield, 2017.

Luhmann, Niklas. "What Is Communication?" *Communication Theory* 2 (1992): 251–59.

Luna, Lynnette. "Coming to America." *Telephony* 239, no. 23 (2000): 10–11.

Lynch, Grahame, and Robert Clark. "Is Plain Old Cellular Enough for Wireless Data?" *America's Network*, July 15, 2000.

Maamria, Kamel. "New Era for Japan." *Telecommunications International* 35, no. 9 (2001): 84–88.

Mailland, Julien, and Kevin Driscoll. *Minitel: Welcome to the Internet.* Cambridge, Mass.: MIT Press, 2017.

Manabe, Noriko. "Going Mobile: The Mobile Internet, Ringtones, and the Music Market in Japan." In *Internationalizing Internet Studies: Beyond Anglophone Paradigms,* edited by Gerard Goggin and Mark McLelland, 316–32. London: Routledge, 2009.

Manovich, Lev. *The Language of New Media.* Cambridge, Mass.: MIT Press, 2001.

Manovich, Lev. *Software Takes Command.* New York: Bloomsbury, 2013.

Marazzi, Christian. *Capital and Language: From the New Economy to the War Economy.* Translated by Gregory Conti. Los Angeles: Semiotext(e), 2008.

Maruchimedia Sofuto Shinkō Kyōkai. *Maruchimedia hakusho 1993* [Multimedia white paper 1993]. Tokyo: Maruchimedia Sofuto Shinkō Kyōkai, 1993.

Maruchimedia Sofuto Shinkō Kyōkai. *Maruchimedia hakusho 1996* [Multimedia white paper 1996]. Tokyo: Maruchimedia Sofuto Shinkō Kyōkai, 1996.

Matsuda, Misa. "Discourses on *Keitai* in Japan." In *Personal Portable Pedestrian: Mobile Phones in Japanese Life,* edited by Mizuko Ito, Daisuke Okabe, and Misa Matsuda, 19–40. Cambridge, Mass.: MIT Press, 2005.

Matsumoto, Craig. "Bankrupt Media Vision Looks to Sell Software Publishing Unit: Serving the Santa Clara Valley." *Business Journal* 12, no. 19 (1994): 16.

Matsunaga Mari. *The Birth of i-mode: An Analogue Account of the Mobile Internet.* Singapore: Chuang Yi Publishing, 2000.

McGray, Douglas. "Japan's Gross National Cool." *Foreign Policy* 130 (May 1, 2002): 44–54.

McKenna, Christopher D. *The World's Newest Profession: Management Consulting in the Twentieth Century.* Cambridge: Cambridge University Press, 2006.

McLelland, Mark, Haiqing Yu, and Gerard Goggin. "Alternative Histories of Social Media in Japan and China." In *The SAGE Handbook of Social Media,* edited by Jean Burgess, Alice Marwick, and Thomas Poell, 53–68. London: SAGE, 2018.

Meeker, Mary. "Internet Trends 2015." *Kleiner Perkins Caufield Byers.* Accessed Jan. 5, 2017. http://www.kleinerperkins.com.

Megler, Veronika. "The Semi-Walled Garden: Japan's 'i-mode Phenomenon.'" Red-paper 166. IBM, Oct. 2001.

Meyer, Marc H., and Alvin P. Lehnerd. *The Power of Product Platforms: Building Value and Cost Leadership.* New York: Free Press, 1997.

Michael, David C., and Yutaka Mizukoshi. "I-mode Is Coming." *Tele.com* 6, no. 8. (2001): 74.

Micklethwait, John, and Adrian Wooldridge. *The Witch Doctors: Making Sense of the Management Gurus.* New York: Random House, 1996.

Miller, Toby, Nitin Govil, John McMurria, Ting Wang, and Richard Maxwell. *Global Hollywood 2.* Rev. ed. London: British Film Institute, 2004.

Moazed, Alex, and Nicholas L. Johnson. *Modern Monopolies: What It Takes to Dominate the 21st Century Economy.* New York: St. Martin's, 2016.

"Mobage." *Wikipedia.* Accessed Sep. 5, 2017. https://en.wikipedia.org.

Mobile Content Forum. *Keitai hakusho 2006* [Cell phone white paper 2006]. Tokyo: Impress Japan, 2005.

Mobile Content Forum. *Keitai hakusho 2011* [Cell phone white paper 2011]. Tokyo: Impress Japan, 2010.

Mobile Content Forum. *Sumaho hakusho 2015* [Cell phone white paper 2015]. Tokyo: Impress Japan, 2016.

Moore, James E. *The Death of Competition: Leadership and Strategy in the Age of Business Ecosystems.* New York: Harper, 1996.

Moore, James E. "Predators and Prey: A New Ecology of Competition." *Harvard Business Review* 71, no. 3 (1993): 75–86.

Morris-Suzuki, Tessa. *Beyond Computopia: Information, Automation and Democracy in Japan.* New York: Kegan Paul International, 1988.

"Most Popular Asia-Based Mobile Messenger Apps as of 4th Quarter 2016, Based on Number of Monthly Active Users (in Millions)." *Statista,* 2017. Accessed June 17, 2017. https://www.statista.com.

Nagle, Angela. *Kill All Normies: Online Culture Wars from 4Chan and Tumblr to Trump and the Alt-Right.* Winchester: Zero Books, 2017.

Nakano Haruyuki. *Manga shinkaron: Kontensu bijinesu wa manga kara umareru* [Theory of manga evolution: The contents business is born from manga]. Tokyo: Burūsu Intāakushonzu, 2009.

Nakayama, Atsuyo. *The Third Wave of Japanese Games.* Tokyo: PHP Institute, 2015.

Nam Kyung Jin and Kwon Youngsun. "How Important Are Real-Time Communications in TV Watching? A Case of AfreecaTV in Korea." 26th European Regional Conference of the International Telecommunications Society (ITS), Madrid, Spain, June 24–27, 2015. https://www.econstor.eu.

Nanba Tomoko. *Fukakō keiei: Chīmu DeNA no chosen* [Ungainly management: Team DeNA's challenges]. Tokyo: Nikkei, 2013.

Narisetti, Raju, and Greg Steinmetz. "Reed Elsevier Wins Bidding for Lexis/Nexis." *Wall Street Journal,* Oct. 5, 1994, A3.

Natsuno Takeshi. *1 chō yen wo kaseida otoko no shigotojutsu* [The work techniques of the man who made one billion yen]. Tokyo: Kodansha, 2009.

Natsuno, Takeshi. *A-la i-mode.* Tokyo: Nikkei BP, 2002.

Natsuno Takeshi. *Atarimae no senryaku shikō* [Obvious strategic thinking]. Tokyo: Fushosha, 2014.

Natsuno Takeshi. *I-mode Strategy.* Translated by Ruth South McCreery. Chichester: John Wiley and Sons, 2003.

Natsuno Takeshi. *The i-mode Wireless Ecosystem.* Translated by Ruth South McCreery. Chichester: John Wiley and Sons, 2003.

Natsuno Takeshi. *I-mōdo sutorateji* [I-mode strategy]. Tokyo: Nikkei BP, 2000.

Natsuno Takeshi. *iPhone vs. Andoroido: Nihon saigo no shōki wo minagasuna!* [iPhone vs. Android: Don't miss Japan's last chance to win!]. Tokyo: Asuki, 2011.

Natsuno Takeshi. *Keitai no mirai* [The future of the mobile phone]. Tokyo: Diamond, 2006.

Natsuno Takeshi. *Naze daikigyō ga totsuzen tsuburerunoka* [Why do massive companies suddenly go bust?]. Tokyo: PHP Business, 2012.

Negoro Tatsuyuki and Ajiro Satoshi. "Keieigaku ni okeru purattofōmu ron no keifu to kongo no tenbō" [A genealogy of and future outlook on platform theory within management studies]. Waseda Daigaku IT Senryaku Kenkyūjo Working Paper 39. Research Institute of IT & Management, Waseda University, Tokyo, May 2011.

Negoro Tatsuyuki and Ajiro Satoshi. "An Outlook of Platform Theory Research in Business Studies." *Waseda Business and Economic Studies*, no. 48 (2012): 1–29.

Negoro Tatsuyuki, Fujitsū Research Institute, and Waseda Business School, eds. *Purattofōmu bijinesu saizensen* [The frontlines of the platform business]. Tokyo: Shōeisha, 2013.

Negoro Tatsuyuki and Kimura Makoto. *Netto bijinesu no keiei senryaku: Chishiki kōkan to baryūchēn* [Management strategies of net businesses: Knowledge exchange and value chains]. Tokyo: Nikka giren, 1999.

Neves, Joshua, and Bhaskar Sarkar. "Introduction." In *Asian Video Cultures*, edited by Joshua Neves and Bhaskar Sarkar, 1–32. Durham, N.C.: Duke University Press, 2017.

Nieborg, David B. "Crushing Candy: The Free-to-Play Game in Its Connective Commodity Form." *Social Media + Society* 1, no. 2 (2015): 1–12.

Nieborg, David B. "The Political Economy of Platformed Production and Circulation." Paper presented at Concordia University, Apr. 6, 2017.

Nieborg, David B., and Thomas B. Poell. "The Platformization of Cultural Production: Theorizing the Contingent Cultural Commodity." *New Media & Society* (Apr. 25, 2018): 1–18.

Nihon Jōhō Shori Kaihatsu Kyōkai, ed. *Jōhōka hakusho 1993* [Informatization white paper 1993]. Tokyo: Konpyūta Eiji, 1993.

Nihon Jōhō Shori Kaihatsu Kyōkai, ed. *Jōhōka hakusho 1994* [Informatization white paper 1994]. Tokyo: Konpyūta Eiji, 1994.

Nihon Jōhō Shori Kaihatsu Kyōkai, ed. *Jōhōka hakusho 1995* [Informatization white paper 1995]. Tokyo: Konpyūta Eiji, 1995.

Nihon Jōhō Shori Kaihatsu Kyōkai, ed. *Jōhōka hakusho 1996* [Informatization white paper 1996]. Tokyo: Konpyūta Eiji, 1996.

Nihon Jōhō Shori Kaihatsu Kyōkai, ed. *Jōhōka hakusho 1997* [Informatization white paper 1997]. Tokyo: Konpyūta Eiji, 1997.

Nihon Jōhō Shori Kaihatsu Kyōkai, ed. *Jōhōka hakusho 2001* [Informatization white paper 2001]. Tokyo: Konpyūta Eiji, 2001.

Nihon Jōhō Shori Kaihatsu Kyōkai, ed. *Jōhōka hakusho 2002* [Informatization white paper 2002]. Tokyo: Konpyūta Eiji, 2002.

Nikkei BP. "Studies of Hypertextual Information Organization." Accessed June 12, 2014. https://nikkeibp.co.jp/item/books/918400.

Nobeoka Kentarō. *Muruchi purojekuto senryaku: Posuto riin no seihin kaihatsu manejimento* [Multiproject strategies: The post-lean management of product development]. Tokyo: Yuhikaku, 1996.

Noguchi Hisashi. *Kontentsu bijinesu: Media sofuto wo sagase* [Contents business: Hunt for media software!]. Tokyo: Jiji Tsūshinsha, 1995.

Nomura Research Institute. *2015nen no Nihon: Arata na "kaikoku" no jidai e* [Japan in 2015: Toward a new era of "opening Japan to the world"]. Tokyo: Toyo Keizai, 2007.

Normile. Dennis. "Coming to America." *Electronic Business* 27, no. 10 (2001): 69–74.

Nozawa, Shunsuke. "The Gross Face and Virtual Fame: Semiotic Mediation in Japanese Virtual Communication." *First Monday* 17, no. 5 (2012). http://first monday.org.

Obara Kazuhiro. *IT bijinesu no genri* [The principles of the IT business]. Tokyo: NHK Shuppan, 2014.

Obara Kazuhiro. *Za purattofōmu* [The platform]. Tokyo: NHK Shuppan, 2015.

O'Donnell, Casey. "The Nintendo Entertainment System and the 10NES Chip: Carving the Video Game Industry in Silicon." *Games and Culture* 6, no. 1 (2011): 83–100.

O'Donnell, Casey. "Production Protection to Copy(right) Protection: From the 10NES to DVDs." *IEEE Annals of the History of Computing* 31, no. 3 (2009): 54–63.

Ohashi, Kana, Fumitoshi Kato, and Larissa Hjorth. "Digital Genealogies: Understanding Social Mobile Media LINE in the Role of Japanese Families." *Social Media + Society* 3, no. 2 (2017): 1–12.

Okada Tomoyuki. "Youth Culture and the Shaping of Japanese Mobile Media: Personalization and the Keitai Internet as Multimedia." In *Personal Portable Pedestrian: Mobile Phones in Japanese Life,* edited by Ito Mizuko, Daisuke Okabe, and Misa Matsuda, 41–60. Cambridge, Mass.: MIT Press, 2005.

"110 Amazing WeChat Statistics and Facts." *DMR.* Accessed Aug. 12, 2018. https:// expandedramblings.com/index.php/wechat-statistics/.

Ōtsuka Eiji. "Kigyō ni kanri sareru saiteki na posutomodan no tame no essei" [An essay for the postmodern optimization of control by companies]. *Sai Zen Sen,* May 17, 2014. http://sai-zen-sen.jp/editors/blog/sekaizatsuwa/otsuka-%20essay .html.

Ōtsuka Eiji. *Media mix ka suru nihon* [The media mixization of Japan]. Tokyo: East Shinsho, 2014.

Ōtsuka Eiji. *Monogatari shōhiron: Bikkuriman no shinwagaku* [A theory of narrative consumption: Mythology of Bikkuriman]. Tokyo: Shin'yōsha, 1989.

Ōtsuka Eiji. *Monogatari shōhiron kai* [A theory of narrative consumption revisited]. Tokyo: Asukī media wākusu, 2012.

Ōtsuka Eiji. *Monogatari shōmetsu ron: Kyarakutāka suru watashi ideorogīka suru monogatari* [A theory of the destruction of narrative: On the characterization of self, and the ideologization of narrative]. Tokyo: Kadokawa, 2004.

Ōtsuka Eiji. *"Otaku" no seishin-shi: 1980-nendai-ron* [A history of the Otaku mind: On the 1980s]. Tokyo: Kōdansha gendai shinsho, 2004.

Ōtsuka Eiji. *Teihon monogatari shōhiron* [A theory of narrative consumption: Standard edition]. Tokyo: Kadokawa, 2001.

Ōtsuka Eiji. "World and Variation: The Reproduction and Consumption of Narrative." Translated by Marc Steinberg. *Mechademia* 5 (2010): 99–116.

Ōtsuka Eiji and Azuma Hiroki. *Riaru no yukue: Otaku/otaku wa dô ikiruka* [The whereabouts of the real: Where are the Otaku/otaku?]. Tokyo: Kodansha, 2008.

Oyama, Shinji. "Japanese Creative Industries in Globalization." In *Routledge Handbook of New Media in Asia,* edited by Larissa Hjorth and Olivia Khoo, 322–32. Abington: Routledge, 2016.

Parker, Geoffrey, and Marshall W. Van Alstyne. "Information Complements, Substitutes, and Strategic Product Design." In *Proceedings of the Twenty-First International Conference on Information Systems,* 13–15. Atlanta: Association for Information Systems, 2000.

Parker, Geoffrey, and Marshall W. Van Alstyne. "Two-Sided Network Effects: A Theory of Information Product Design." *Management Science* 51, no. 10 (2005): 1494–504.

Parker, Geoffrey G., Marshall W. Van Alstyne, and Sangeet P. Choudary. *Platform Revolution: How Networked Markets Are Transforming the Economy and How to Make Them Work for You.* New York: W. W. Norton, 2016.

Partner, Simon. *Assembled in Japan: Electrical Goods and the Making of the Japanese Consumer.* Berkeley: University of California Press, 1999.

Perzanowski, Aaron, and Jason Schultz. *The End of Ownership: Personal Property in the Digital Economy.* Cambridge, Mass.: MIT Press, 2016.

Plantin, J.-C., Paul N. Edwards, Carl Lagoze, and C. Sandvig. "Infrastructure Studies Meet Platform Studies in the Age of Google and Facebook." *New Media & Society* 20, no. 1 (2016): 293–310.

Prigg, Mark. "Could Brown Bear and Cony the Bunny Replace Emoji? Hi-Tech Characters Based on Messaging 'Stickers' Set to Launch in US." *Daily Mail,* Mar. 20, 2015.

Ratliff, John. "NTT Docomo and Its i-mode Success." *California Management Review* 44, no. 3 (2002): 55–71.

Richardson, Adrian. "Mapping Out the Chat App Landscape." Nexmo blog, June 12, 2015. https://www.nexmo.com.

"Riding the Wave, Carefully." *J@pan Inc,* Oct. 2001. http://www.japaninc.com.

Rifkin, Jeremy. *The Age of Access: The New Culture of Hypercapitalism, Where All of Life Is a Paid-For Experience.* New York: Penguin, 2001.

Rochet, Jean-Charles, and Jean Tirole. "Platform Competition in Two-Sided Markets." *Journal of the European Economic Association* 1, no. 4 (2003): 990–1029.

Rohlfs, Jeffrey. "A Theory of Interdependent Demand for a Communications Service." *Bell Journal of Economics and Management Science* 5, no. 1 (1974): 16–37.

Rohwer, Jim, et al. "On Top of the World." *Fortune* 142, no. 9 (2000): 164–76.

Roquet, Paul. *Ambient Media: Japanese Atmospheres of Self*. Minneapolis: University of Minnesota Press, 2016.

Rose, Frank. "Keyword: Context." *Wired*, Dec. 1, 1996. http://www.wired.com.

Rysman, Marc. "The Economics of Two-Sided Markets." *Journal of Economic Perspectives* 23, no. 3 (2009): 125–43.

Safire, William. "The Summer of This Content." *New York Times*, Aug. 9, 1998, SM18.

Saito Kumiko. "Magic, Shōho, and Metamorphosis: Magical Girl Anime and the Challenges of Changing Gender Identities in Japanese Society." *Journal of Asian Studies* 73, no. 1 (2014): 143–64.

Sakai, Naoki. *Translation and Subjectivity: On "Japan" and Cultural Nationalism*. Minneapolis: University of Minnesota Press, 1997.

Salazkina, Masha. "Introduction: Film Theory in the Age of Neoliberal Globalization." *Framework* 56, no. 2 (2015): 325–49.

Sasaki Atsushi. *Nippon no shisō* [Japanese thought]. Tokyo: Kodansha, 2009.

Sasaki Toshinao. *Netto mirai chizu* [Net future map]. Tokyo: Bungei Shunjū, 2007.

Sasaki Toshinao. *Niconico dōga ga mirai wo tsukuru: Dwango monogatari* [Niconico video is creating the future: The Dwango story]. Tokyo: Ascii Media Works, 2009.

Satō Kichinosuke. *Subete wa koko kara hajimaru: Kadokawa Grūpu wa nani wo mezasu ka* [Everything starts from here: What Kadokawa Group is aiming for]. Tokyo: Kadokawa Group Holdings, 2007.

Satō Tatsuo. "Address to the Company." *Kadokawa K-Net* 86 (2013): 2–4.

Saxenian, AnnaLee. *The New Argonauts: Regional Advantage in a Global Economy*. Cambridge, Mass.: Harvard University Press, 2006.

Scheible, Jeff. *Digital Shift: The Cultural Logic of Business*. Minneapolis: University of Minnesota Press, 2015.

Schmalensee, Richard. "Jeffrey Rohlfs' 1974 Model of Facebook: An Introduction." *Competitive Policy International* 7, no. 1 (2011): 301–38.

Schmidt, Eric, and Jonathan Rosenberg. *How Google Works*. New York: Grand Central Publishing, 2014.

Schoenberger, Chana R. "Queen of Mobile." *Forbes*, May 9, 2008. https://www.forbes.com.

Scholtz, Trebor. *Uberworked and Underpaid: How Workers Are Disrupting the Digital Economy*. Cambridge: Polity, 2016.

Seave, Ava, Bruce Greenwald, and Jonathan A. Knee. *The Curse of the Mogul: What's Wrong with the World's Leading Media Companies*. New York: Portfolio/Penguin, 2009.

Sekizawa Tadashi. "Nikkei sangyō shinbun sōkan 20 shūnen tokubetsu kōgikai: Fūjitsū shachō Sekizawa Tadashi-shi" [On the occasion of the 20th anniversary of the founding of Nikkei Sangyō newspaper: A special lecture by Fujitsū president Mr. Sekizawa Tadashi]. *Nikkei Shinbun*, Oct. 7, 1993, 5.

Shapiro, Carl, and Hal Varian. *Information Rules: A Strategic Guide to the Network Economy*. Boston: Harvard Business School Press, 1999.

Sheff, David. *Game Over: How Nintendo Zapped an Industry, Captured Your Dollars, and Enslaved Your Children.* New York: Random House, 1993.

Shiba Tomonari. *Hatsune Miku wa naze sekai wo kaeta no ka?* [Why did Hatsune Miku change the world?]. Tokyo: Ōta, 2014.

Shimizu Kinnichi. *Fujitsū no maruchi media bijinesu: Infura kara kontentsu made no jigyō senryaku* [Fujitsū's multimedia business: Business strategies from infrastructure to contents]. Tokyo: Oasis Shuppan, 1995.

Shimizu Kinnichi. *Fujitsū no saibā bijinesu* [Fujitsū's cyber business]. Tokyo: Ōesu Shuppan, 1998.

Shin Mukoeng and Ha Jonggi. *Yabai LINE* [Dangerous LINE]. Tokyo: Kobunsha, 2015.

Simon, Phil. *The Age of the Platform: How Amazon, Apple, Facebook, and Google Have Redefined Business.* Henderson, Nev.: Motion, 2011.

Smythe, Dallas W. "On the Audience Commodity and Its Work." In *Media and Cultural Studies: Keyworks,* edited by Meenakshi Gigi Durham and Douglas Kellner, 230–56. London: Blackwell, 2001.

Snickars, Pelle, and Patrick Vonderau, eds. *Moving Data: The iPhone and the Future of Media.* New York: Columbia University Press, 2012.

Srnicek, Nick. *Platform Capitalism.* Malden, Mass: Polity, 2016.

Standage, Tom. "The Internet, Untethered: A Survey of the Mobile Internet." *The Economist,* Oct. 11, 2001, 3–20.

Steinberg, Marc. *Anime's Media Mix: Franchising Toys and Characters in Japan.* Minneapolis: University of Minnesota Press, 2012.

Steinberg, Marc, and Alexander Zahlten, eds. *Media Theory in Japan.* Durham, N.C.: Duke University Press, 2017.

Steinbock, Don. *Wireless Horizon: Strategy and Competition in the Worldwide Mobile Marketplace.* New York: AMACOM, 2003.

Stewart, Alex. "Why Do Cybirds Suddenly Appear?" *J@pan Inc,* Aug. 2003. http://www.japaninc.com.

Stone, Brad. *The Everything Store: Jeff Bezos and the Age of Amazon.* New York: Little, Brown, 2013.

Stone, Brad. *The Upstarts: How Uber, Airbnb, and the Killer Companies of the New Silicon Valley Are Changing the World.* New York: Little, Brown, 2017.

Stross, Randall. *Planet Google: One Company's Audacious Plan to Organize Everything We Know.* New York: Free Press, 2008.

Suarez, Fernando F., and Michael A. Cusumano. "The Role of Services in Platform Markets." In *Platforms, Markets and Innovation,* edited by Annabelle Gawer, 77–98. Cheltenham: Edward Elgar, 2009.

"Summary of Consolidated Financial Results for the Three Months Ended March 31, 2017." LINE, Apr. 26, 2017. https://scdn.line-apps.com/stf/linecorp/en/ir/all/17Q1QuarterlyReport.pdf.

Tanaka Emi. "Media mikkusu no sangyō kōzō" [The industry structure of the media mix]. In *Kontentsu sangyōron* [On the contents industry], edited by

Deguchi Hiroshi, Tanaka Hideyuki, and Koyama Yūsuke, 159–188. Tokyo: Tokyo Daigaku Shuppan, 2009.

Tapscott, Don. *The Digital Economy: Promise and Peril in the Age of Networked Intelligence.* New York: McGraw-Hill, 1996.

"TEKTRONIX; Announces Redirection of Graphic Workstation Development." *Business Wire,* Aug. 22, 1985.

"Tencent Investor Kit: Fact Sheet." *Tencent,* Mar. 23, 2016. http://www.tencent.com/en-us/content/ir/fs/attachments/investorintro.pdf.

Terranova, Tiziana. "Free Labor: Producing Culture for the Digital Economy." *Social Text* 18, no. 2 (2000): 33–58.

"The Times in Numbers 2017." *New York Times,* n.d. https://www.nytco.com/the-times-in-numbers-2017/.

Tiwana, Amrit. *Platform Ecosystems: Aligning Architecture, Governance, and Strategy.* Amsterdam: Elsevier, 2014.

"Top Sites in Japan." *Alexa.* Accessed Aug. 26, 2018. http://www.alexa.com.

"Top Sites Ranking for All Categories in Korea, Republic Of." *SimilarWeb.* Accessed June 17, 2017. https://www.similarweb.com.

Tsing, Anna. "Supply Chains and the Human Condition." *Rethinking Marxism: A Journal of Economics, Culture & Society* 21, no. 2 (2009): 148–76.

"2015 Search Engine Market Share by Country." *Return on Now.* Accessed June 17, 2017. http://returnonnow.com.

van Dijck, José. *The Culture of Connectivity: A Critical History of Social Media.* Oxford: Oxford University Press, 2013.

van Dijck, José, Thomas Poell, and Martijn de Waal. *The Platform Society.* Oxford: Oxford University Press, 2018.

Vincent, James. "99.6 Percent of New Smartphones Run Android or iOS." *The Verge,* Feb. 16, 2017. https://theverge.com.

Vogelstein, Fred. *Dogfight: How Apple and Google Went to War and Started a Revolution.* New York: Farrar, Straus and Giroux, 2013.

Vonderau, Patrick. "The Spotify Effect: Digital Distribution and Financial Growth." *Television and New Media* (Nov. 21, 2017): 1–17.

Warbelow, Art, and Jirō Kokurō. "AUCNET TV Auction Network System." 1989. Harvard Business School case study 9-190.

"WeChat Daily Active Users Reached 570 Mln in Sep 2015." *China Internet Watch,* Nov. 5, 2015. http://www.chinainternetwatch.com.

Wheelwright, Steven C., and Kim B. Clark. "Creating Project Plans to Focus Product Development." *Harvard Business Review* 70, no. 2 (1992): 70–82.

Wheelwright, Steven C., and Kim B. Clark. *Revolutionizing Product Development: Quantum Leaps in Speed, Efficiency, and Quality.* New York: Free Press, 1992.

Whittaker, Jason. *The Cyberspace Handbook.* London: Routledge, 2004.

Willis, Susan. *A Primer for Everyday Life.* London: Routledge, 1991.

Wollen, Peter. "Godard and Counter-Cinema." In *Narrative, Apparatus, Ideology,* edited by Philip Rosen, 120–29. New York: Columbia University Press, 1986.

Womack, James P., Daniel T. Jones, and Daniel Roos. *The Machine That Changed the World.* 1990. Reprint, Cambridge, Mass.: MIT Press, 2007.

World Auto Trade: Current Trends and Structural Problems; Hearings before the Subcommittee on Trade of the Committee on Ways and Means, House of Representatives, Ninety-Sixth Congress, Second Session, March 7, 18, 1980. Washington, D.C.: U.S. Govt. Print. Off., 1980.

"The World's Most Valuable Resource Is No Longer Oil, but Data." *The Economist,* May 6, 2017.

Woyke, Elizabeth. *The Smartphone: Anatomy of an Industry.* New York: New Press, 2014. Kindle version.

Wu, Tim. *The Attention Merchants: The Epic Scramble to Get inside Our Heads.* New York: Vintage, 2016.

Wu, Tim. *The Master Switch: The Rise and Fall of Information Empires.* New York: Random House, 2010.

Yamakawa Satoru. *Jirei de wakaru monogatari māketingu* [Narrative marketing understood through examples]. Tokyo: Nihon nōritsu kyōkai manajimento sentā, 2007.

Yano, Christine. *Pink Globalization: Hello Kitty's Trek across the Pacific.* Durham, N.C.: Duke University Press, 2013.

Ynostra, Roger. "Whose Content Is It Anyway?" *Graphic Arts Monthly* 67 (1995): 12.

Yoda, Tomiko. "Girlscape: The Marketing of Mediatic Ambience." In *Media Theory in Japan,* edited by Marc Steinberg and Alexander Zahlten, 173–99. Durham, N.C.: Duke University Press, 2017.

Young, Deborah. "Settling Up." *Wireless Review* 18, no. 9 (2001): 24–27.

Yutaka Mizukoshi, Kimihide Okino, and Olivier Tardy. "Lessons from Japan." *Telephony* 240, no. 3 (2001): 92–96.

Zahlten, Alexander. "1980s Nyū Aca: (Non)Media Theory as Romantic Performance." In *Media Theory in Japan,* edited by Marc Steinberg and Alexander Zahlten, 200–20. Durham, N.C.: Duke University Press, 2017.

Zahlten, Alexander. *The End of Japanese Cinema: Industrial Genres, National Times, and Media Ecologies.* Durham, N.C.: Duke University Press, 2018.

Zhen, Pei, and Lionel M. Ni. *Smart Phone and Next Generation Mobile Computing.* San Francisco: Elsevier, 2006.

Zittrain, Jonathan. *The Future of the Internet and How to Stop It.* New Haven, Conn.: Yale University Press, 2008.

Zöllner, Reinhard, and Nakamura Yoshio, eds. *Culture and Contents: Understanding Contents Business in Japan and the World.* Munich: Iudicium, 2010.

Index

MARC STEINBERG is associate professor of film studies at Concordia University, Montreal. He is author of *Anime's Media Mix: Franchising Toys and Characters in Japan* (Minnesota, 2012) and its Japanese expanded version and translation, *Naze Nihon wa "media mikkusu suru kuni" nano ka?* (Why is Japan a "media mixing nation"?). He is coeditor of *Media Theory in Japan.*